Adverse Reproductive Outcomes in Families of Atomic Veterans: The Feasibility of Epidemiologic Studies

Committee to Study the Feasibility of, and Need for,
Epidemiologic Studies of Adverse Reproductive Outcomes
in the Families of Atomic Veterans

Medical Follow-up Agency

INSTITUTE OF MEDICINE

NATIONAL ACADEMY PRESS
Washington, D.C. 1995

National Academy Press • 2101 Constitution Avenue, N.W. • Washington, D.C. 20418

NOTICE: The project that is the subject of this report was approved by the Governing Board of the National Research Council, whose members are drawn from the councils of the National Academy of Sciences, the National Academy of Engineering, and the Institute of Medicine. The members of the committee responsible for the report were chosen for their special competencies and with regard for appropriate balance.

This report has been reviewed by a group other than the authors according to procedures approved by a Report Review Committee consisting of members of the National Academy of Sciences, the National Academy of Engineering, and the Institute of Medicine.

The Institute of Medicine was chartered in 1970 by the National Academy of Sciences to enlist distinguished members of the appropriate professions in the examination of policy matters pertaining to the health of the public. In this, the Institute acts under both the Academy's 1863 congressional charter responsibility to be an adviser to the federal government and its own initiative in identifying issues of medical care, research, and education. Dr. Kenneth I. Shine is president of the Institute of Medicine.

Support for this project was provided by the Department of Veterans Affairs (contract no. V101-(93)P-1469).

International Standard Book No. 0-309-05340-4

Additional copies of this report are available in limited quantities from:

National Academy Press
2101 Constitution Avenue, N.W.
P.O. Box 285
Washington, D.C. 20055

Call 800-624-6242 or 202-334-3313 (in the Washington Metropolitan Area)

B669

The serpent has been a symbol of long life, healing, and knowledge among almost all cultures and religions since the beginning of recorded history. The serpent adopted as a logotype by the Institute of Medicine is a relief carving from ancient Greece, now held by the Staatlichemuseen in Berlin.

iii

Preface

At the request of the Department of Veterans Affairs and mandated in Public Law 103-446, Section 508, enacted on November 2, 1994, the Medical Follow-up Agency (MFUA) of the Institute of Medicine (IOM) established a committee to review the available data and scientific literature on the health effects of exposure to ionizing radiation and to prepare a report on the feasibility of studying veterans exposed to ionizing radiation and the risk of health effects in their spouses, children, and grandchildren. Specifically, the committee, which was established in January 1995, was asked to address the following three questions:

1. Is it feasible to conduct an epidemiologic study to determine whether there is an increased risk of adverse reproductive outcomes in the spouses and of adverse health effects in the children and grandchildren of veterans who participated in atmospheric weapons tests, served with the occupation forces of Hiroshima or Nagasaki, Japan, prior to July 1, 1946, or were prisoners of war of Japan with an opportunity for exposure to ionizing radiation similar to that of the occupation forces (the Atomic Veterans)?

2. If such a study is feasible, how much time and money would be required to organize and implement it?

3. Are there other sources of information that would yield similar results at lower cost or in less time?

The committee met in Washington, D.C., on three separate occasions, January 23–24, March 2–3, and April 17–18, 1995. At the first meeting the committee solicited comments—oral, written, or both—from members of the public and, in particular, representatives of the various concerned veterans groups. Comments were also sought from members of and spokespersons for the House and Senate Committees on Veterans Affairs and the U.S. Department of Veterans Affairs. This report sets forth the committee's assessment of the feasibility of studies of adverse reproductive outcomes in families of servicemen exposed to ionizing radiation.

The committee is deeply appreciative of the comments and written submissions of the various concerned individuals and groups, and of the work of the staff of the Medical Follow-up Agency. In particular, we are indebted to Richard Miller, J. Christopher Johnson, John Zimbrick, Erin Bell, and Pamela Ramey-McCray for staff support. We thank Michael Hayes for editorial review.

William J. Schull, *Chair*

Contents

Adverse Reproductive Outcomes in Families of Atomic Veterans: The Feasibility of Epidemiologic Studies

Executive Summary

Over the past several decades, public concern over exposure to ionizing radiation has increased. This concern has manifested itself in different ways depending on the perception of risk to different individuals and groups within the population of the United States and the circumstances of their exposure. One such group is made up of those servicemen who participated in the atmospheric testing of nuclear weapons at the Nevada Test Site or in the Pacific Proving Grounds, who were involved in cleanup activities in Hiroshima or Nagasaki in the fall of 1945 and the spring of 1946, or who were prisoners of war who may have been assigned duties in those cities at the times of the bombings or shortly thereafter. Initially, this concern focused on the veterans themselves and may have been stimulated by early reports of an excess incidence of leukemia among participants in the 1957 Nevada test shot known as SMOKY (Caldwell et al., 1980, 1983). The Radiation-Exposed Veterans Compensation Act of 1988 (Public Law 100-321) recognized this concern and identified 13 cancers (specifically, leukemia, multiple myeloma, lymphoma except Hodgkin's disease, and cancers of the thyroid, breast, pharynx, esophagus, stomach, small intestine, pancreas, bile ducts, gall bladder, and liver) that were deemed to be presumptively service connected, and thus compensable. In 1994, this act was amended to include two additional sites of cancer, namely, the salivary gland and the urinary tract (Public Law 102-578).

Now the concern of some people extends beyond the health risk to the veteran and involves health issues related to their children, grandchildren, and spouses. As a result of these concerns the U.S. Congress, in Public Law 103-

1

446, Section 508, directed the Secretary of Veterans Affairs (VA) to enter into an agreement with the Medical Follow-up Agency (MFUA) of the Institute of Medicine (IOM) to convene a panel of appropriate individuals to carry out the following tasks:

> An evaluation of the feasibility of a study to determine the nature and extent, if any, of the relationship between the exposure of veterans to ionizing radiation and the occurrence of (1) genetic defects and illness in their children and grandchildren, (2) adverse reproductive outcomes experienced by their wives, and (3) periparturient diseases of the mother that are the direct result of such adverse reproductive outcomes.

The task of the committee, as elaborated by the VA, was to address the following three questions:

> 1. Is it feasible to conduct an epidemiologic study to determine whether there is an increased risk of adverse reproductive outcomes in the spouses and of adverse health effects in the children and grandchildren of Atomic Veterans?
> 2. If such a study is feasible, approximately how much time and money would be required to organize and implement it?
> 3. Are there other sources of information that would yield similar results at a lower cost or in less time?

CONCLUSIONS

The committee has addressed each of these questions, with the conclusions summarized here. The background information and rationale that served as a basis for these findings are described in detail in the full report.

1. Is it feasible to conduct an epidemiologic study to determine whether there is an increased risk of adverse reproductive outcomes in the spouses and of adverse health effects in the children and grandchildren of Atomic Veterans?

Conclusion: The committee's assessment is that there are insurmountable difficulties in finding and contacting a sufficiently large number of study subjects (offspring of Atomic Veterans), in establishing an accurate measure of dose for each veteran, in detecting the extremely small potential risk at low doses, in identifying and reliably documenting reproductive outcomes over a 50–year interval, and in the measuring of other factors that have been observed to cause reproductive problems, and therefore, might confound any observed relationship between radiation exposure and reproductive problems. These difficulties become even greater in the grandchildren of these veterans. The committee concluded, therefore, that, as a result of the difficulties enumerated above, the cohort of Atomic Veterans does not provide a practical opportunity for a scientifically adequate and epidemiologically valid study.

2. If such a study is feasible, approximately how much time and money would be required to organize and implement it?

Conclusion: Since the committee does not believe that an epidemiologic study is feasible, it did not consider in detail the time and money that would be required. However, on the basis of past and current studies of radiation-exposed cohorts, the committee estimates that such a study would cost tens of millions of dollars and would last at least a decade.

3. Are there other sources of information that would yield similar results at a lower cost or in less time?

While experimental animal studies could address some of the scientific issues discussed in this report, the committee has interpreted this charge to pertain to alternative epidemiologic studies that could yield similar results at a lower cost or in less time.

Conclusion: The committee suggests some studies that might be informative, but notes that these too will have limitations. Commonly, these limitations are related to sample size, population composition, uncertainty of dose, the presence of concurrent disease, and other confounding factors. Although studies of these groups may have their own merits, the committee concludes that they may not adequately address the immediate concerns of the Atomic Veterans.

ORGANIZATION AND FRAMEWORK OF THE REPORT

To evaluate the feasibility of conducting an epidemiologic study of the families of Atomic Veterans, the committee felt it necessary to begin with a review of the fundamental principles of epidemiology, radiation biology, and genetics. This review is then followed by discussions of current information on the risk of genetic mutations due to environmental exposure, definitions and possible causes of adverse reproductive outcomes, the factors to be considered when determining the feasibility of a study, and finally, a review of possible alternative approaches for evaluating the health effects of exposure to low levels of ionizing radiation.

ADVERSE REPRODUCTIVE OUTCOMES

Adverse reproductive outcomes include such endpoints as the inability to conceive (sterility or infertility), the premature spontaneous termination of a pregnancy (spontaneous abortion), the birth of an infant with a congenital malformation or with mental and physical retardation, and premature death (stillbirth, neonatal or infant death). These events are not rare in the general population. For example, the committee estimates that 15,000 children with major birth defects would be expected among the 500,000 or so offspring of the 210,000 Atomic Veterans in the absence of any radiation effects. Many host and environmental factors contribute to the origin of such outcomes. These outcomes may arise through maternally or paternally derived inherited defects, through exposure to noxious environmental agents, including ionizing radiation, smoking, or the consumption of alcohol, through preexisting maternal illness (such as diabetes) or illness during pregnancy, and through poor nutrition. The presence of so many causes of adverse reproductive outcomes makes it difficult to design an epidemiologically valid study and to know the cause of the health outcome in any particular person.

The likelihood that genetic effects may be seen following exposure of the human male to ionizing radiation is an important factor in assessing the feasibility of studying adverse reproductive outcomes among Atomic Veterans and their families. The human data on which an estimate of this likelihood can be based are limited, and much rests on animal studies. The role of paternal exposure to radiation or chemicals has not been investigated extensively in humans (Olshan and Faustman, 1993). However, some effect of chemicals or radiation on the sperm, chromosome, and fertility has been demonstrated (Wyrobek et al., 1983; Martin et al., 1986; Geneseca et al., 1990). Thus, male-mediated environmental exposures may affect pregnancy outcome. Nonetheless, it should be emphasized that most of the findings are based on a small number of studies, and the paucity of information and definitive mechanistic models make interpretations of the available data arguable.

ESTIMATION OF GENETIC RISK

Three observations need to be understood when estimating genetic risk:

1. The results of the mutagenic effects of ionizing radiation are indistinguishable phenotypically from those caused by other agents or those that arise spontaneously.

2. If a study shows statistically that an increase in an adverse health outcome in a population may be due to additional radiation exposure, this fact cannot be taken as proof that the illness of any particular individual is a result of exposure to radiation.

3. Relatively high doses of radiation (greater than 2,000 mSv [200 rem]) would add only a small number of additional cases of genetic disorders to the large number that are expected to occur as a result of spontaneous mutations, most of which have existed in the population for many generations.

Although much of what is known about the mutagenic effects of exposure to ionizing radiation rests on animal studies, notably those involving the mouse, two sets of human data have played important roles in estimating the risks of ionizing radiation to human populations: (1) data pertaining to the estimation of the spontaneous burden of genetic disease and disability (the number of cases normally present in the population) and (2) data on the Hiroshima-Nagasaki atomic bomb survivors and their children conceived subsequent to parental exposure that have been used to estimate the doubling dose.

Overall, the studies of health outcomes (listed in the report) in the Hiroshima-Nagasaki population have revealed a small but statistically nonsignificant increase in negative health outcomes, including congenital defects, stillbirths, non-cancer mortality and some chromosomal abnormalities, among the children of the atomic bomb survivors conceived after the bombing compared with the outcomes among the children of nonexposed parents. This small increase is consistent with what would be predicted from studies of ionizing radiation in experimental animals and provides the best currently available basis for estimating the doubling dose in human beings (Neel and Schull, 1991).

FEASIBILITY OF AN EPIDEMIOLOGIC STUDY

The feasibility of a study of adverse reproductive outcomes among the families of veterans exposed to ionizing radiation hinges largely on the answers to four questions: (1) How is a suitable sample or cohort of exposed persons affected among the total at risk (numerators and denominators) to be defined, and can this be done without inadvertently introducing selection biases that could obscure a true effect or produce a spurious one? (2) Will that sample or cohort be large enough to reveal effects of the magnitude anticipated on the basis of present knowledge? (3) What is the probable dose distribution among the members of that sample or cohort, and how reliable are the individual dose estimates? (4) What approaches are available for identifying adverse reproductive outcomes accurately and completely? Each of these questions is considered separately.

Question 1. Definition of the Study Sample

Anecdotal information can be valuable in establishing the need for an epidemiologic study, but self-volunteered information is unlikely to provide a basis for reliable estimates of risk since experience shows that persons with a personal interest in an exposure to some hazard are more likely to respond. Accordingly, a scientifically defensible and valid study of the effects of ionizing radiation on adverse reproductive outcomes depends on the availability of a representative sample of exposed veterans and their families, and the means to establish these outcomes without reference to whether they are normal or abnormal. The Nuclear Test Personnel (NTPR) program of the Defense Nuclear Agency (DNA) has identified some 210,000 veterans who participated in one or more atmospheric tests involving the detonation of a nuclear weapon. These individuals or a suitably large and representative sample might provide the basis for a study cohort, and it seems probable that deaths among these veterans could be determined through the records of the VA or other sources. However, it is far more difficult to trace an unbiased sample of living persons, given the lack of identifying information in the original records. Furthermore, the available records do not contain information on the reproductive histories of the veterans (that is, their children), estimated for the purposes of this report to be about 500,000 in number. For reasons described elsewhere in this report, difficulties in locating offspring and obtaining such information reliably and accurately at this late date appear to be insurmountable. Thus, the committee concludes that, whereas a study of the life status and health problems of the Atomic Veterans themselves is feasible (and is in fact being done), the means do not exist to obtain information on adverse reproductive outcomes among their children and grandchildren in a manner complete enough for an unbiased estimate of the risk, if any, stemming from exposure to ionizing radiation.

Question 2. Sample Size

The committee has approached the issue of sample size by posing two related but separate questions. First, the committee asked, If current estimates of the probable risk of adverse health effects among the children of the exposed fathers are correct, how large a sample would be needed to demonstrate that risk? Second, the committee asked, If sample size is fixed, how large a risk would have to exist to be statistically demonstrable? In the first instance, to calculate the sample size required, the committee estimated what the maximum relative risk would be, given the data on the effects of ionizing radiation provided by animal and human studies. The *maximum* relative risk, for the purposes of determining sample size, was estimated to be 1.002 (0.2 percent increase in adverse health effects in the exposed population compared with the adverse health effects in an unexposed one). By using the maximum relative risk of 1.002 (equivalent to an excess relative risk of 0.002) and given that the expected frequency of major congenital defects in the general population observable at birth is about 3 percent, a sample size of 212,000,000 children would be necessary to demonstrate a statistically significant increased risk among the children of exposed veterans compared with that among the children of nonexposed fathers. In the second instance, assuming a sample size of 500,000 children and a frequency of major congenital defects observable at birth of 3 percent, the minimum detectable relative risk among the children of the most heavily exposed veterans is estimated to be 1.3, which is equivalent to an excess relative risk of 0.300. This excess risk, it will be noted, is 150 times larger than the maximum estimated excess relative risk based on current evidence, that is, 0.002. The committee concludes, therefore, that the sample size requirements are too great for a valid epidemiological study of adverse health effects among the children of Atomic Veterans to be performed.

Question 3. Dose

A crucial issue in assessing the feasibility of a study of adverse reproductive outcomes among the families of Atomic Veterans is the matter of dose. The demonstration that a particular endpoint increases in frequency as dose increases is a compelling argument for causality. If the individual doses are, on average, low and, moreover, unreliable, the demonstration of such a trend is unlikely. Inspection of the estimated doses received by the Atomic Veterans indicates that the majority probably received less than 5 mSv (0.5 rem). Everyone is exposed to background radiation that can vary in intensity depending upon location. Roughly 90% of the Atomic Veterans have estimated doses below the extra background radiation received from living in Denver, CO, or in an area of similar altitude, compared to a more typical background area in the United States. Thus, without detailed residential histories the movement of these veterans in their post-service years could be an important confounder in the estimation of their cumulative doses. Moreover, medical exposure to ionizing radiation in the United States is very common. While the typical effective dose (NCRP, 1987) from a chest x-ray is small—approximately 0.06 mSv (0.006 rem)—other diagnostic procedures such as an upper gastrointestinal examination (2.45 mSv, 0.245 rem), barium enema (4.05 mSv, 0.405 rem), or computed tomography (CT scan) (1.1 mSv, 0.110 rem) result in significantly higher doses. Thus, the majority of Atomic Veterans would have received exposures to some tissues very similar to those that occur in certain medical examinations, and in the absence of detailed information on diagnostic irradiation, this, too, poses a possible source of error in the estimation of their total doses. The Committee to Study the Mortality of Military Personnel Present at Atmospheric Tests of Nuclear Weapons (IOM, 1995) concluded that the existing dose information on Atomic Veterans is inadequate to estimate individual doses with the requisite consistency to support an epidemiologic study. Taken together, these concerns make it highly improbable that a valid study is possible.

Question 4. Identification of Adverse Reproductive Outcomes

Study of reproductive outcomes among the families of Atomic Veterans requires being able to identify both normal and abnormal outcomes in an unbiased manner. A cohort study—identifying groups of veterans who are similar with the exception of exposure and following them over time to determine if the rates of reproductive outcomes differ by exposure group—would seem to be a logical approach. However, this is more difficult than it would first appear. These groups are likely to have completed their families at least 15 years ago, and experience has shown that the hospital records necessary to identify adverse reproductive outcomes during a period of from 15 to 50 years ago are not likely to be available at a quantity and of a quality sufficient for the purposes of an epidemiologic study. In addition, when one considers health outcomes that are not readily diagnosed at birth, such as learning disabilities and mental retardation, the challenges of finding documentation become even greater. As a result, these latter health endpoints are difficult to study epidemiologically in defined contemporary populations and would be extremely difficult, if not impossible, to study adequately in a historical cohort.

The potential for recall bias is of particular concern when studying health problems for which individuals may not routinely receive medical attention, such as spontaneous abortions, infertility, and developmental delays. Spontaneous abortions present a number of methodological problems even for studies in contemporary populations (Sever, 1989). Unbiased studies of spontaneous abortions in populations that were at the height of their reproductive lives more than 30 years ago would not be possible. In groups of women who have been questioned about their history of spontaneous abortion, recall seems to be relatively accurate for the period up to 20 years prior to interview; however, before that time, recall is poor on the basis of a comparison of contemporary reports with later recall (Wilcox and Horney, 1984). This is in the absence of any concern about a potential association with an exposure that might lead to reporting or recall bias (White et al., 1989).

ALTERNATIVE APPROACHES

Data on the occurrence of adverse reproductive outcomes following exposure to ionizing radiation could be derived from a variety of cohorts in addition to the atomic bomb survivors, such as the children of (1) people residing in areas where the background of naturally occurring radiation is substantially higher than usual, (2) individuals, other than the Atomic Veterans, exposed to fallout from atmospheric weapons testing, (3) people living near nuclear installations, (4) individuals exposed occupationally, (5) patients undergoing medical diagnostic procedures, and (6) patients undergoing medical therapy for benign or malignant disease. Each of these cohorts has strengths and limitations in sample size, population composition, adequate dose information, the presence of concurrent diseases, and the presence of confounding factors. The committee recognizes the real concerns of the Atomic Veterans as expressed by their representatives, but it must conclude that epidemiologic studies cannot adequately address these concerns.

1

Introduction

Over the past several decades, public concern over exposure to ionizing radiation has increased. This concern has manifested itself in different ways depending on the perception of risk to different individuals and groups within the population of the United States and the circumstances of their exposure. One such group is made up of servicemen who participated in the atmospheric testing of nuclear weapons at the Nevada Test Site or in the Pacific Proving Grounds, who were involved in cleanup activities in Hiroshima or Nagasaki in the fall of 1945 and the spring of 1946, or were prisoners of war who may have been assigned work duties in those cities at the times of the bombings or shortly thereafter. Initially, this concern focused on the veterans themselves and may have been stimulated by early reports of an excess incidence of leukemia among participants in the 1957 Nevada test shot known as SMOKY (Caldwell et al., 1983). The Radiation-Exposed Veterans Compensation Act of 1988 (Public Law 100-321) recognized this concern and identified 13 cancers (specifically, leukemia, multiple myeloma, lymphoma except Hodgkin's disease, and cancers of the thyroid, breast, pharynx, esophagus, stomach, small intestine, pancreas, bile ducts, gall bladder, and liver) that were deemed presumptively service connected, and thus compensable. In 1994 this act was amended to include two additional sites of cancer, namely, the salivary gland and urinary tract (Public Law 102-578).

Now the concern of some people extends beyond the health risk to the veterans and involves health issues related to their children, grandchildren, and spouses. As a result of the concerns expressed by the spokespeople for these veterans, their spouses, and children, the Committees on Veterans Affairs of the House of Representatives and the Senate in Public Law 103-446, Section 508, Study of Health Consequences for Family Members of Atomic Veterans of Exposure of Atomic Veterans to Ionizing Radiation, directed the Secretary of Veterans Affairs (VA) to enter into an agreement with the Medical Follow-up Agency (MFUA) of the Institute of Medicine (IOM) to convene a panel of appropriate individuals to carry out the following:

1. An evaluation of the feasibility of a study to determine the nature and extent, if any, of the relationship between the exposure of veterans to ionizing radiation and the occurrence of (1) genetic defects and illness in their children and grandchildren, (2) untoward pregnancy outcomes experienced by their wives, and (3) periparturient diseases of the mother which are the direct result of such untoward pregnancy outcomes.

2. If such a study is feasible, the committee was asked to estimate how much time and money would be required to organize and implement it.

3. Finally, the committee was asked to determine if other sources of information would yield similar results at a lower cost or in less time (while experimental animal studies could address some of the scientific issues discussed in this report, the committee has interpreted this charge to pertain to alternative epidemiologic studies that could yield similar results at a lower cost or in less time).

The panel was directed to submit its evaluation to the Secretary not later than 180 days after the date of enactment of Public Law 103-446.

The veterans covered under this law include (1) any serviceman who was exposed (as determined by the Secretary) to ionizing radiation as a result of (a) participation while on active duty in the Armed Forces in an atmospheric nuclear test that included detonation of a nuclear device, or (b) served in the Armed Forces with the United States occupation force in Hiroshima or Nagasaki, Japan, before July 1, 1946, or (c) was interned or detained as a prisoner of war in Japan before that date in circumstances providing the opportunity for exposure to ionizing radiation comparable to the exposure of individuals who served with such occupation force before that date, and (2) any other veteran who the Secretary designates for coverage under the study (Public Law 103-446).

For the committee's evaluation of the issues implicit in its charge to be thorough, it was necessary that its members represent a broad array of disciplines. Accordingly, members were chosen to represent expertise in radiation dosimetry, epidemiology, ethics, genetics, radiation biology, reproductive biology, teratology, and statistics. To discharge its responsibilities, the committee has met monthly since it was constituted.

AIMS OF THIS REPORT

Through its deliberations, review of relevant documents, understanding of issues relative to conducting epidemiologic studies, and discussion of various study strategies, the committee answered the questions established under Public Law 103-446, Section 508. The committee evaluated the feasibility of a study to determine the nature and extent of the relationship between the exposures of defined groups of veterans to ionizing radiation and selected health effects on their children, grandchildren, and wives. To determine the feasibility of an epidemiologic study, it is necessary to establish which health endpoints are thought to be associated with a particular exposure or exposures. Although Public Law 103-446, Section 508, describes the general categories of health effects that are to be evaluated, to meet its charge, it was important for the committee to obtain information from the Atomic Veterans regarding their specific concerns.

The testimony provided by the Atomic Veterans and their representatives gave the committee a framework for the determination of the scope of the health effects to be considered. Since the Atomic Veterans have identified a variety of potential health effects among their wives, children, and grandchildren as being of concern, the committee had to conduct a thorough evaluation of the feasibility of studying diverse health effects occurring over a number of years in a geographically dispersed population.

The overarching aim of the committee's deliberations was therefore to search for a workable approach to addressing these concerns. This search entailed evaluation of feasibility considerations in two general areas. The first area related to the logistic aspects of conducting a scientifically valid epidemiologic study. Here, the issue of feasibility revolved around the definition and assessment of exposures, confounding factors, and outcomes and hinged on the availability and quality of the data. These were largely practical matters that must be considered in determining whether an epidemiologic study that would meet the requirements for good epidemiologic research could be conducted. Ancillary to that are questions regarding the ethics of conducting studies if their potential ability to address the concerns of affected individuals is limited.

The second general area of feasibility is the scientific background that supports the conduct of a study. Feasibility concerns here include the evidence on which one would base estimates of the anticipated magnitude of increased risk. This led the committee to the consideration of three additional areas: (1) the evidence for a biological basis and mechanisms for the effects; (2) the increase in risk that might be anticipated and the statistical power of a study to identify those levels of increased risk, if present; and (3) the potential contribution of factors other than the exposure of interest to any apparent increase in risk. The aims of the committee included consideration of all of the above issues in a search for a feasible approach to addressing the concerns raised by the Atomic Veterans and their representatives.

In addition to evaluating the feasibility of an epidemiologic study, the committee was asked to estimate how much time and money would be required to conduct such a study, if one was deemed to be feasible. It is important to note that such an estimate is predicated not only on determining feasibility but also on determining the scope necessary for a study to address the concerns of the veterans and their families. The type of study design and the data sources to be used also play a key role in determining how much time and money would be required. For example, a case-control study based on accessing, reviewing, and abstracting existing records would be much less expensive than a cohort study that required tracking subjects over time and conducting interviews for primary data collection. Cost and time are also affected by the numbers and types of outcomes to be studied, and in reviewing the outcomes included in Public Law 103-446, Section 508, it is not clear that all of them would have equal weight.

The third charge to the committee was to determine whether there are other sources of information that could yield similar results at a lower cost or in less time.

2

Basic Epidemiologic Issues

The diseases resulting from exposure to ionizing radiation are usually indistinguishable from the diseases that occur in the general population. Leukemia looks the same in an atomic bomb survivor as it does in a person who was not exposed to an atomic bomb blast. Only the increased frequency in an exposed group indicates that radiation exposure may have played a role. However, increased frequency alone is usually not enough to establish a causal relationship between an exposure and a health effect. To help evaluate causality it is also necessary to consider the following questions (Hill, 1965):

- Does the frequency of the disease increase as the dose increases (that is, is there a dose-response relationship)?
- Can the findings be duplicated by other investigators?
- Is a similar effect seen in experimentally exposed animals (some species may not be susceptible) and other laboratory studies?
- Can alternative explanations, such as cigarette smoking and heredity be excluded?
- Is the finding biologically plausible?
- Does the effect cease to occur when the cause is removed?

The history of the discovery of an intrauterine effect of radiation is illustrative. In the 1920s physicians began to note that mothers of infants with small

13

head size and mental retardation had often received radiotherapy early in pregnancy (Murphy, 1928). Within a few years these case reports led to a U.S. mail survey of these occurrences, well documented in hospitals, that doubled the known number of cases to 30 (Goldstein and Murphy, 1929). Soon after, the same effect was induced experimentally in rats, and extensive mouse studies demonstrated a close relationship between developmental stage at exposure and effect observed. This finding was confirmed by the study of Japanese atomic bomb survivors exposed in utero (Plummer, 1952; Miller, 1956; Yamazaki and Schull, 1990), in whom the incidence of the effect could be determined. They were from a defined population exposed to a variety of doses and could be studied as a cohort, that is a group exposed together, in this instance, at a single moment in time, for the full spectrum of diseases that occur over the entire life span. The results confirmed those in the earlier case series. In addition, a dose–response relationship that could not be attributed to other variables such as malnutrition was found, the effect was biologically plausible, and the incidence returned to normal in children born subsequently to those who had been exposed in utero.

Small head size is a teratogenic effect (induced by exposure of the developing fetus to the teratogen in contrast to a genetic effect, which is passed from parent to child through a defective gene. Suspicion that radiation was responsible for this effect began with a number of case reports. Similar disease cluster investigations have led to much of what is known about environmental causes of cancer or birth defects (Miller, 1978). Clusters are generally defined as aggregations of events in space and time. They pose a challenge because they are often based on small numbers, are epidemiologically complex since a multitude of potential risk factors need to be considered, require comparisons with background rates that are often difficult to obtain, and involve statistical and mapping techniques that are controversial (Rothenburg et al., 1990). For example, the study of more than 100 cancer clusters by experts at the U.S. Public Health Service's Centers for Disease Control from 1961 to 1982 revealed no environmental causes (Caldwell, 1990) for these clusters. The reason is that almost all small clusters occur by chance, even for such rare diseases as childhood leukemia. When a health professional, a parent, or anyone else notices a group of cases in a small area, there is the tendency to draw boundaries of time and geography tightly around the cases. As a result, the occurrence of disease appears to be unusually high in this defined area. Cancer clusters occur by chance continuously throughout the United States (Neutra et al., 1990). When it seems an environmental exposure might be responsible, experienced investigators may conduct a case–control study to determine what in the environment is suspect. In such a study, the histories of affected individuals ("cases") are compared with those of similar unaffected persons ("controls") to determine if the exposure in question was significantly more frequent among cases than among controls. However, the elements necessary to help establish causality outlined above must also be evaluated. If the environmental exposure has not been found to cause the cancer or some other disease after heavy exposures, as in industry, after accidental exposures, or after receiving medications, it is not plausible to expect an effect at low doses.

Family clusters of disease are often genetic in origin, but for noninfectious diseases, family clusters of disease are rarely due to a shared environment. Important heritable cancers, such as retinoblastoma in children (Knudson, 1988) and diverse forms of cancers that occur before 45 years of age in individuals with Li-Fraumeni syndrome (Li et al., 1988), have been identified through the study of such family clusters. Clinical identification of these rare disorders has led to new understanding of the genesis of many common cancers (Levine, 1995). The study of unusual family cancer clusters due to genetics has been especially rewarding. Because cancer occurs so often in the population at large, family clusters most commonly occur by chance. Subtle exposures such as diet are often suspected, but they are difficult to establish.

The interpretation of clusters of adverse pregnancy outcomes are particularly problematic to public health officials, epidemiologists, and biostatisticians. The same difficulties described above would apply to an even greater extent when assessing multiple adverse reproductive events within a family. Guidelines for cluster investigations have recently been set forth by the Centers for Disease Control (1990).

Any study that focuses on a single potential risk factor for disease needs to consider the established and probable risk factors as potentially confounding variables. Possible interactions among risk factors are often difficult to interpret because they may occur by chance in studies that consider a large number of factors. Selection bias can occur if, for example, the participants in the study differ from the control group in aspects other than the factor under study. Such bias can affect the generalizability of the findings and, more importantly, measures of association (Kelsey et al., 1986).

The foregoing illustrates the epidemiologic approach to investigating the cause(s) of disease. One function of epidemiology is to measure disease frequency, which can be expressed as prevalence (the number of affected people in the population at a given time) or incidence (the number of new cases in a given time interval). These studies, when properly done, provide numerators (the number of people affected) and denominators (the number of people at risk), which are the bases of estimating risk.

One of the most important considerations in interpreting epidemiologic data is the strength or magnitude of an association between exposure and disease. A number of terms are used to express strength of association or risk, and it is often very difficult to compare one epidemiologic study with another because of the different terms and types of analysis used. The most commonly used term for cohort studies is relative risk (RR) or excess relative risk (ERR). They both are an expression of the risk of the exposed group relative to that of some nonexposed group. A relative risk of 1.0 means that the exposed group has the same risk as the control group, and implies an excess relative risk of 0. If there is a relative risk of 2, then the exposed population would be twice as likely as a nonexposed control group to develop a condition as a result of exposure. This equates to a doubling of the risk. The control group must be essentially the same age, sex, and so forth, as the exposed group or these variables should be considered in the analysis.

An increased relative risk does not necessarily mean that there has been an effect of a given exposure. For example, if two populations are being compared and the control group for some reason has a lower disease incidence or mortality rate than would normally be expected, the relative risk will be greater than 1.0 even though there is no excess incidence of disease or mortality in the exposed population. For this reason an increased relative risk or excess relative risk should not be viewed as establishing causality until such factors have been clarified.

Typically, the estimation of risk is accompanied by the assessment of random variation. This is accomplished through the use of statistical significance testing. Using an appropriate test for the type of data at hand, the investigator will derive the *p*-value, which is the *probability* that an effect as extreme as that observed could have occurred by chance, given that there is, in fact, no relationship or association between exposure and disease (the null hypothesis). By convention, if the *p*-value is less than or equal to 0.05, then the association between exposure and disease is considered to be statistically significant. This means that there is no more than a 5 percent or a 1-in-20 probability of observing a result as extreme as that observed because of chance alone if there is, in fact, no association. If the *p*-value is greater than 0.05, then the effect is considered to be not statistically significant. The current recommended practice in medical and epidemiologic research is also to report an informative measure—the confidence interval. The confidence interval represents the range of possible values for the parameter of interest (e.g., the relative risk) that is consistent with the observed data within specified limits. The width of the confidence interval reflects the sample size, in that the narrower the interval, the less variability in the estimate of the effect measured; likewise, the greater the width, the greater the variability.

A potential problem in epidemiologic studies that sometimes goes unrecognized is bias of reporting or in selecting study participants. A mailed questionnaire may draw a disproportionately higher number of responses from exposed individuals who think that they have been injured as a result of their exposure than from unexposed individuals or those who do not believe that they have been injured as a result of their exposure. Similarly, investigators may publish positive results but may not publish negative results, and scientific journals generally favor reports of positive results. Consequently, the public is often left with a distorted view of scientific reality. It warrants noting, too, that case reports and case series commonly result in false interpretations rather than new insights into etiology.

3

Feasibility and Design of an Epidemiologic Study

Epidemiology is the art of the practical. Unlike the researcher in a laboratory experiment, who has control over the relevant variables, including the genetic homogeneity of the animals and the precise exposures for an experiment, the epidemiologist must deal with events that have already occurred. Thus, the feasibility and design of an epidemiologic study are dictated by external, practical matters as well as by the hypothesis under question. Some practical concerns are:

1. availability of an appropriate study population;
2. size and composition of the study population;
3. completeness (and lack of bias) with which study subjects can be enrolled;
4. magnitude and distribution of exposure to the hazard being studied;
5. accuracy with which the exposure can be measured (measurement of absorbed dose, as in the atomic bomb survivors, is extremely important since the most compelling evidence of causality is the demonstration of a dose–response relationship);
6. accuracy of disease identification (history of disease should be confirmed by hospital records, and causes of death should be determined by obtaining copies of death certificates);
7. background rate of the disease being studied;

8. expected increase in disease among the exposed group; and
9. availability of information on other factors that might determine disease.

STUDY COHORT

One starting point for any epidemiologic study is a defined population. Either samples of this group of people or the entire group is studied. If only a sample is studied, it is important that the sample be representative of the whole, or at least that the differences between the sample and the total population be clearly identified. Problems arise when the study sample differs from the population in ways that are ill-defined or unknown. This may happen when people in the study population cannot be traced, have died, or decline to participate. The higher the percentage of the population lost for these reasons, the less certain are the conclusions that can be drawn from the people actually studied.

MEASUREMENT OF EXPOSURE

Hazardous exposures are difficult to measure under the best of circumstances. When the study takes place years after the exposure has occurred, the difficulties are even greater. Few exposures can be measured accurately in retrospect. Specific problems of radiation dose measurements are discussed in Chapter 8.

The less precisely an exposure is measured, the harder it is to find a clear association between exposure and disease. When measurement of exposure is poor, the high-dose group may be so diluted by less exposed people that no effect can be seen. In addition, exposures to other disease-causing factors (such as cigarette smoking) may confound the connection between the suspected exposure and the disease. Information on these other factors (known as confounders) must be collected in an epidemiologic study. However, it can be as difficult to collect information on an individual's confounding factors for previous time periods as it is to measure the suspected exposure under study.

DEFINING THE DISEASE AND
ASCERTAINING THE CASES

In an ideal study, the disease would be defined by examination of pathologic tissues or other direct diagnostic means. More often one must rely on medical records or self-reports. Medical records may be incomplete or inconsistent because of the lack of standardized methods for examination and diagnosis. Physicians usually do not arrive at a diagnosis by a well-standardized routine but by experience and judgment, which may vary widely among individuals. Also, medical records are so decentralized that it can be very difficult to find individual records. Records are usually saved for a maximum of 20 years; this time period is often less if hospitals close or doctors end their practice.

Reproductive problems are especially hard for epidemiologists to study, and the various types of reproductive outcomes require different kinds of ascertainment strategies. Hospital records may be good for identifying problems of delivery, but poor for finding fertility problems. No single epidemiologic design can identify all of the types of reproductive endpoints. This means that some choices regarding which endpoints are most important must be made at the onset of a study. The specific reproductive endpoints that should be included in a study of radiation-exposed men would depend on biologic plausibility, prior research results, and to some degree, anecdotal information. A detailed discussion of radiation and its most likely reproductive effects is provided in Chapter 6.

A further difficulty in studying reproductive endpoints is that many are never detected or recorded in the medical system at all. Reproductive problems are not always obvious, and not all are diagnosed correctly. At least half of U.S. couples who are infertile never take their problem to a doctor (OTA, 1988). Most spontaneous abortions occur so early in pregnancy that the woman does not know that she was pregnant (Wilcox et al., 1988). Even serious birth defects can be underreported in routine records (Lie et al., 1994). When records are inadequate, the couples themselves must be relied upon to provide information. Self-reports have their own limitations. People must sometimes be asked to recall events that occurred many years earlier. Study subjects are known to have trouble recalling even such major events as a spontaneous abortion, much less the clinical details of those events (Wilcox and Horney, 1984), and men typically recall medical events relating to pregnancy, delivery, infancy, and childhood less well than women.

Once the criteria for what defines a case are established, then the cases must be located or ascertained. In addition, every person in the reference population should be contacted personally by an investigator. This is usually impractical and may even be impossible. Existing medical records can be useful, but they are often difficult to track down, especially in the United States where people may get their medical care from many sources. Self-identification of cases may be possible, but then the concern is that people with a personal or even a financial interest in an exposure to some hazard will selectively respond. When reproductive difficulties are the topic, there is the added problem that the most seriously affected individuals (such as a baby with a major birth defect) may no longer be alive, making ascertainment and diagnosis more difficult.

Ultimately, however, the feasibility of an epidemiologic study depends on satisfying not one but a series of scientific requirements. It is necessary to evaluate the power of the proposed study (which will depend on the population size of those exposed and the magnitude of the expected risk differences) and to assess whether the necessary data on health effects, exposure, and potential confounders can be obtained (refer to Chapter 9).

Epidemiologic studies pose two closely related ethical issues: privacy and confidentiality. Privacy in this context refers to keeping particularly sensitive information about oneself a secret, whereas confidentiality refers more generally to keeping personal data out of the hands of others without the authorization of the subjects. Epidemiologic studies usually involve a review of various types of existing records, including medical records, and may also involve interviews with individuals regarding medical and other personal information. It has become standard ethical practice in the United States to have epidemiologic protocols reviewed by federally mandated Institutional Review Boards (IRB) to ensure that researchers

> take adequate steps to preserve the confidentiality of the data they collect, requiring that they specify who will have access to the data, how and at what point in the research personal information will be separated from the data, and whether the data will be retained at the conclusion of the study. IRB reviewers also make sure that the informed consent of the subjects will be obtained before interviews are conducted . . . (Wallace, 1982; OPRR, 1993).

There is some discretion on the part of IRBs to authorize record review by accountable individuals who agree to protect confidentiality, at least when no information that can be linked to the individual is kept. However, when confidential information can be linked to the individual, each subject must consent to the study under most circumstances (OPRR, 1993). It is also important to have a mechanism to inform participants of the results of the study and to counsel them when appropriate, especially if the results might influence their health care, medical future, or other important life decisions.

4

Basic Principles of Radiation Biology

To understand how ionizing radiation can damage biologic systems, it is necessary to understand what ionizing radiation is and how it interacts with tissues in the body. There are two types of ionizing radiation: nonparticulate (gamma and X rays) and particulate (alpha and beta particles, neutrons and protons). Both forms can transfer energy into a substance. If the energy is high enough, the incoming radiation may eject electrons from atoms along its path through the material. This process is called *ionization*.

The composition of ionizing radiation determines how it interacts with the matter surrounding it. Electromagnetic radiation is a form of light energy. The electromagnetic spectrum extends from very long wavelengths, which include electric power, television and radio, to those in the middle, which include visible and ultraviolet light. Approaching the other end of the spectrum, those with the shorter wavelengths include microwaves, radar, and infrared radiation, and as the wavelength becomes very short, the spectrum contains the highly energetic waves of ionizing radiation. The particulate types of radiation consist of subatomic particles that may be either charged or neutral and can vary considerably in size and mass.

Not all types of radiation are equally penetrating, and the depth that a particular radiation penetrates into material depends on the energy and the type of radiation. Almost all types of ionizing radiation are much more easily stopped by dense material (such as lead) than by water or tissue in the human body. In general, X rays and gamma rays are more penetrating than the particulate types of radiation such as beta and alpha particles.

Beta particles are electrons and typically penetrate into the tissue only a centimeter or so. Their limited range means that they can damage internal organs only when ingested or inhaled, but they can be an external hazard to exposed skin if they are present in sufficient concentrations. Alpha particles are much larger and heavier than beta particles and have a greater electrical charge. This makes it even more difficult for them to penetrate tissue. A typical alpha particle from radioactive materials, such as plutonium, will not even penetrate the external dead layer of skin tissue. Radioactive materials that emit alpha particles are a hazard only if they are inhaled or ingested and get into the cells of the body in sufficiently large concentrations. Because X rays and gamma rays travel as very-high-energy electromagnetic waves, they can penetrate the human body quite easily. Either external or internal sources of gamma radiation can be hazardous to the whole body because of the extraordinary penetrating ability of the radiation that they emit.

Radiation damage to genetic material can occur directly or indirectly when ionizing radiation passes through the nucleus of a cell. For direct damage to occur the radiation must hit genetic material. Since the volume of the sensitive material is so very small compared with the total volume of the cell and its surrounding tissue, the probability of that happening is remote. If the radiation interacts in close proximity to the genetic material, the interaction can create a free radical that can then drift close enough to the DNA to damage it.

The vast majority of those types of radiations that do interact produce ionization and, subsequently, free radicals. These free radicals generally will recombine in microseconds with no biologic effect. Even if they do not recombine, it must be remembered that only a very small portion of the cell is represented by genetic material and that the diffusion distance of free radicals is very short. Thus, most free radicals are not able to interact with genetic material. Furthermore, it is common for electromagnetic radiation to pass through a cell without interacting with the cell or its contents.

It is also clear from radiobiological research that even if there is an interaction with a segment of genetic material as a result of the presence of ionizing radiation, the cell possesses many repair mechanisms. These ensure that few of the genetic interactions will result in an adverse health outcome. This can be understood more easily if one thinks about the number of ionizing events that occur in each person daily as a result of natural background radiation. Approximately 25 million ionizing events occur within the body of each person each hour of each day. Since people are usually healthy, these ionizing events rarely lead to mutations or obvious harm.

Radiation is measured and described in a number of ways. One can use a meter or other device to measure radiation in air, that is, exposure. The units used to express exposure are either roentgens or coulombs per kilogram. This measurement method applies only to ionizing electromagnetic radiation, such as gamma ray and X rays, not to particulate radiation. Also, since there are differences between the levels of penetration of different types of radiation in tissue as well as differences in the distribution of energy along the path of the ionization, a more useful expression is the energy actually deposited in a certain amount of tissue. This measurement is referred to as *absorbed dose*. The unit of absorbed dose is either the gray or the rad. One gray equals 100 rads. However, measurement of the energy deposited in tissue does not account for all of the differences in biologic effects between different radiation types.

This fact is important because the spatial distributions of ionization in material for gamma rays, beta particles, and alpha particles are different. Alpha particles interact very readily with the matter that they penetrate. They are called high-linear-energy-transfer (high-LET) radiation because they dissipate their energy rapidly, producing very short, dense tracks of ionization. Because of their high-LET characteristics alpha particles can be much more damaging, for a given absorbed dose, than low-LET radiations such as beta particles and gamma rays. Low-LET radiations ionize the atoms in their paths much less frequently and produce tracks that are much less densely ionized.

It is possible to compare the biologic effects from different types of radiation by using radiation weighting factors. The factor for alpha particles is about 20 and that for gamma and beta radiation is approximately 1, indicating that it takes about 20 times more gamma or beta radiation than alpha radiation to cause a given effect. The dosimetry measurement that allows the differences in biologic effectiveness of various types of radiation to be combined is called the *equivalent dose*. It is calculated by multiplying the absorbed dose by the radiation weighting factors. The unit of equivalent dose is the sievert or the rem. One sievert (Sv) equals 100 rem.

BIOLOGICAL EFFECTS

There are two general types of biological effects from ionizing radiation: deterministic effects and stochastic effects. Stochastic effects are those effects whose *frequency* in the exposed population is a direct function of dose, no matter how low the dose is; these effects are commonly regarded as having no threshold. Deterministic effects are those effects whose *severity* in the exposed individual is dependent on dose; these effects are commonly regarded as having a threshold. Deterministic effects are often the result of cell killing. Since in most organs and tissues there is a continuous process of loss and replacement of cells, a slight increase in the rate of loss due to cell killing can be compensated for by an increase in the replacement rate. If the radiation exposure is higher, there may be some reduction in function of that particular tissue.

For most healthy individuals, the probability of causing harm because of deterministic effects will be close to zero at absorbed doses of less than 100 mSv

(10 rem) (NRC, 1990). Some tissues are much more resistant than others to cell killing, and no effects are demonstrated until absorbed doses are in the range of several sieverts (several hundred rem). A notable exception is the sensitivity of the testes during germ cell formation. For deterministic effects, there is a practical threshold below which the body is able to compensate with cellular replacement. If doses are high enough and involve exposure to the entire body, then death will occur. In the absence of medical treatment, an acute (brief) whole-body dose of 3,500 mSv (350 rem) will result in the deaths of approximately half of the exposed individuals.

At small increments of dose above the level of background radiation, the probability of inducing either an additional cancer or a genetic defect is negligible, and the number of cases of cancer or genetic effects attributable to a small increase in dose in a very large exposed group may well be less than one. Although there may be no definable threshold, epidemiologic studies show that, as the radiation exposure becomes lower, the magnitude of any effect in a population is so small that it cannot be identified against the background of spontaneously occurring cancer or genetic effects. Scientific studies of the 86,000 atomic bomb survivors from Hiroshima and Nagasaki showed that 37,800 individuals died from all causes. About 8,000 persons died from cancer but the excess cancer cases due to radiation were estimated to be fewer than 450 during the entire 40-year follow-up period (Mettler and Upton, 1995).

The incidence and severity of many radiation-induced biologic effects are a function not only of the level of dose but also of the rate at which the radiation is received. A simple explanation for this is that a given radiation dose that is spread over time allows the body to use repair mechanisms, whereas very high doses given in a very short time may overcome the body's ability to use repair mechanisms. The human data that contribute to current estimates of radiation effects are based on high-dose and high-dose-rate exposures. In general, a dose and/or a dose rate effectiveness factor (DDREF) is applied to high-dose/high-dose-rate estimates to assess biologic effects in those receiving low dose rates or low doses. A DDREF between 2 and 10 is found in experiments in animals. However, for most radiation protection purposes, a conservative factor of 2 in reducing the expected effect is used. For most of the Atomic Veterans exposed to fallout, the dose rate would generally be low, whereas for those exposed directly to a weapon at the time of explosion, the dose rate would be high. Given the dose information, presented in Chapter 9, it would appear that almost all of the Atomic Veterans have what would be classified as a low dose, and therefore, a reduction factor of 2 for potential biologic effects could be assumed, that is, half that expected at a high dose and a high dose rate.

SOURCES OF RADIATION EXPOSURE

It is important to place the magnitude of exposure received by the Atomic Veterans in perspective. Exposure to ionizing radiation comes from two major sources: natural (background) radiation and technology-induced radiation, often referred to as manmade radiation. In most, if not all, countries, natural sources of radiation constitute the major source of radiation exposure for the population, with the next largest source being medical applications.

In the United States the average annual effective dose of naturally occurring background radiation is about 3 mSv (0.30 rem) per year (NCRP, 1987). Of this, about 2 mSv (0.20 rem) comes from exposure to radon, 0.28 mSv (0.028 rem) from cosmic rays, 0.39 mSv (0.039 rem) comes from naturally occurring nuclides in the human body, and finally, 0.28 mSv (0.028 rem) comes from naturally occurring radioactive materials within the ground.

There can be significant variations in the levels of background radiation even within the United States. For example, the natural background radiation from cosmic rays and terrestrial sources in Denver, Colorado, is 50 percent higher than the national average (NCRP, 1987). Natural background exposure during 70 years of a lifetime is an effective dose of approximately 200 mSv (20 rem). If one lived in Denver or in an area of equivalent altitude, one's lifetime effective dose would be approximately 20 mSv (2 rem) higher than the national average. An inspection of the doses received by the Atomic Veterans indicates that the majority received less than 5 mSv (0.5 rem). Only 10 percent of the Atomic Veterans appear to have received doses that would exceed the naturally occurring difference in radiation resulting from living in Denver compared with that from living in an area in the United States with a more typical level of background radiation.

Medical exposure to ionizing radiation is very common in the United States. Typical effective doses (NCRP, 1987) from a chest X ray are approximately 0.06 mSv (0.006 rem) but other procedures such as an upper gastrointestinal examination (2.45 mSv; 0.245 rem), barium enema (4.05 mSv; 0.405 rem), or computed tomography (CT) scan (1.1 mSv; 0.11 rem) result in significantly higher doses. Medical radiation procedures avoid unnecessary exposure to the gonads, which keeps doses of genetic importance below the doses given above. The National Council on Radiation Protection and Measurements (NCRP) estimates that the *annual* genetically significant dose (GSD) from medical exposures received by the general population is in the range of 0.2–0.3 mSv (0.02–0.03 rem) (NCRP, 1987). Thus, the majority of Atomic Veterans would have received exposures to some tissues very similar to those that occur as a result of standard medical examinations, and in the absence of detailed information on diagnostic irradiation this, too, poses a possible source of error in the estimation of their possible doses.

POTENTIALLY SENSITIVE SUBGROUPS

The committee has reviewed the scientific literature for evidence of subgroups of the population potentially sensitive to ionizing radiation. Two groups of individuals are known to have genetic or chromosomal defects and have increased sensitivities to various types of ionizing radiation. The most notable are individuals with ataxia-telangiectasia (AT), a rare inherited disorder (2 or 3 per 100,000 live births) in which children have a staggering gait (ataxia), bloodshot eyes (conjunctival telangiectasia), chromosomal breakage on culture of their fibroblasts, and a high risk of lymphoma. When the lymphoma is treated with conventional doses of X rays, a severe, often lethal acute radiation reaction occurs. The abnormality in patients with AT is a result of cell killing because of their inability to repair DNA damaged by ionizing radiation. An extensive search has been made of other diseases with a defect in DNA repair capacity that might influence radiosensitivity. In five rare single-gene disorders, some impairment of survival of fibroblasts was found in culture after gamma irradiation, but not to the same degree as that found in AT homozygotes. The AT heterozygotes have normal test results (Paterson et al., 1984).

5

Genetic Principles and Issues

A major underlying concern at issue is whether the exposure to ionizing radiation could cause an excess occurrence of genetic disease in the offspring of the Atomic Veterans. There is an intuitive sense in the use of the term genetic disease that is understood by both scientists and the general public for most purposes, but the meaning of the term is, in fact, quite elusive when it is needed for the purpose of considering the feasibility of an investigation. Many diseases, in the end, probably result from an interaction of inborn (heritable) susceptibility and a lifelong exposure to various environmental factors.

Classically, genetic disease embraces three categories: cytogenetic (chromosomal), single gene, and multifactorial. To be specific, there are some syndromes of multiple malformations that have chromosomal abnormalities (cytogenetic syndromes), such as Down syndrome (which is due to an extra chromosome number 21). Other diseases are due to defects in single genes and are inherited according to the laws of Mendel, such as familial hypercholesterolemia, Huntington disease, and neurofibromatosis. Two major patterns of Mendelian inheritance are dominant (disease is seen when one copy of the disease causing gene from one parent is present) and recessive (disease occurs only when two copies of the disease causing genes, one derived from each of the two parents, are present). Finally, many diseases and birth defects, such as diabetes, cancer, infertility, cleft palate or spina bifida, are multifactorial or polygenic (no major gene or single environmental agent is the cause of the disease or birth defect). Some birth defects have major environmental determinants, which are called teratogens. Examples include rubella (German measles), which acts directly on the developing human being to cause cataracts, and ionizing radiation, which causes small head size and mental retardation from birth.

The deoxyribonucleic acid (DNA) in each human cell is organized into 46 separate packages called chromosomes that can be seen when the cell is dividing, because the chromosomes are condensed into tight particles. Most body cells, called somatic cells, have two sets of chromosomes, one from each parent, for a total of 46 chromosomes in each cell. Sperm and egg cells, which are called germ cells, have just one set of chromosomes. A female has two X chromosomes and a male has one X and one Y chromosome; hence, the X and Y chromosomes are called sex chromosomes. The remaining 22 chromosomes are called autosomes.

In the normal somatic cell, each gene has two versions, one from the mother (on the chromosome present at fertilization from the egg, the maternal germ cell) and one from the father (on the chromosome from the sperm, the paternal germ cell). Usually, the two genes work in concert to produce normal cellular structures and functions. Sometimes, one or both genes may be so altered that a disease may result.

A mutation is a sudden and permanent change in the DNA sequence. Most mutations may have no known effect, but some are harmful and contribute to the occurrence of human disease. Many mutations arise spontaneously, but they may also be caused by exposure to certain chemicals and ionizing radiation.

A mutation may change only a single gene (referred to as a point mutation), or may affect the integrity of the chromosome in a manner that involves more than one gene; even the latter type of genetic lesion may not be detectable under the microscope. Ionizing radiation appears to be a poor point mutagen. Many radiation-related mutations are believed to be chromosomal in nature, often involving the deletion of a portion of the chromosome.

6

Current Knowledge and Estimation of Genetic Risk

The estimation of the genetic effect of ionizing radiation on human populations has been a matter of concern since World War II. The two main bodies involved are the United Nations' Scientific Committee on the Effects of Atomic Radiation (UNSCEAR) and the U.S. National Academy of Sciences' Committee on the Biological Effects of Ionizing Radiations (BEIR). In addition, the International Commission on Radiological Protection (ICRP, Oftedal and Searle, 1980) as well as the Nuclear Regulatory Commission (Nuclear Regulatory Commission, 1985) have published documents in which genetic risk estimates are included. All tend to give similar estimates because they all use basically the same set of data.

In what follows it is important to understand three points:

1. The health effects resulting from mutations induced by ionizing radiation are indistinguishable from those resulting from other agents or that arise spontaneously.

2. Even if a significant increase in some endpoint is shown statistically to be due to additional radiation exposure, no specific case can be proved to be ascribable to that exposure.

3. Finally, even high doses of radiation (greater than 2,000 mSv [200 rem]) will add only a small number of additional cases of genetic disorders to the relatively large number that are expected to occur as a result of spontaneous mutations, most of which have existed in the population for many generations.

BASIC ESTIMATION EQUATION

Many of the estimates of the genetic impact of ionizing radiation on human populations have made use, in one form or another, of the following formula:

$$I = S \times 1/DD \times MC \times D , \qquad \text{(Equation 1)}$$

where:

I is the increased number of cases (per generation) of genetic effects due to radiation, often called the induced burden,

S is the number of cases (per generation) normally present in a population not exposed to additional radiation, the spontaneous burden,

DD is the doubling dose (see below),

MC is the mutation component (see below), and,

D is the dose of additional radiation to which the population is exposed.

The use of Equation 1, especially when applied as in the case of the Atomic Veterans, requires some explanation. Let B = the total burden to the population of some genetic disease or class of diseases, e.g. the total number of cases arising per generation; and let m = mutation rate. Now consider the equality:

$$\Delta B/B = \Delta m/m \times (\Delta B/B) / (\Delta m/m). \qquad \text{(Equation 1a)}$$

that is, the relative change in B equals the relative change in m times the relative change in B to the relative change in m.

Suppose that in the dose range being considered, we may assume that mutation rate is a linear function of dose, for example, $m = m_0 + bD$, where m_0 is the spontaneous mutation rate. Then

$$\Delta m/m = (m - m_0) / m_0 = bD / m_0 .$$

But, m_0/b is the doubling dose—exactly that dose that induces m_0 mutations. So we see that $\Delta m/m = D/DD$. Furthermore, $(\Delta B/B) / (\Delta m/m)$ is the mutation component (Crow and Denniston, 1981), the relative change in the burden to the relative change in the mutation rate. Hence, we see that Equation 1a can be written $\Delta B/B = (D/DD) \times MC$ or since $\Delta B = I$ and $B = S$, we have $I = S \times (D/DD) \times MC$, which is Equation 1.

Now, in the usual application of Equation 1, I ($ = \Delta B$) applies to the change in the burden from just before a permanent change in the dose to the time the population reaches the new equilibrium between mutation and selection. But, in fact, the relevant time period is determined by how the mutation component (MC) is defined. The MC may be defined to apply to any number of generations after the change in radiation exposure. In particular, it may be defined to apply to the first generation after the increase in dose. If the increase in dose is permanent, I (= ΔB) slowly increases from generation one to equilibrium; if the dose increase is temporary (e.g., a burst, as in the case of the atomic veterans), then I (= ΔB) increases in the first generation but then slowly decreases until the old equilibrium is reestablished.

The simplest example is that of an autosomal dominant gene. At equilibrium between mutation and selection, the frequency of the trait is 2m/s, where m is the mutation rate and s is the selection coefficient. If the mutation rate increases permanently from m to m(1 + k) then in generation n after the increase the frequency is:

$$2[m/s + km (1 - (1 - s)^n/s)].$$

If the increase in mutation is only a burst, the frequency in generation n is:

$$2[m/s + km (1 - s)^{n-1}].$$

The first frequency eventually rises to 2m (1 + k)/s while the second returns to 2m/s. In generation one, the two are identical. Using the definition of mutation component given above, the mutation component in generation n can be defined as $MC_n = 1 - (1 - s)^n$ in the case of a permanent change and as $MC_n = s(1 - s)^{n-1}$ in the case of the burst. The term "mutation component," without specification of the generation, conventionally refers to MC_∞.

For more complicated traits, no simple formulas exist, but for threshold traits, such as congenital abnormalities, the mutation component in the first generation is generally less than one or two percent. Equation 1 thus can be applied to the case of the Atomic Veterans by using the value of mutation component that applied to the first-generation effect. Subsequent generations (e.g., grandchildren) would show even smaller effects.

If the dose-response curve is not linear but concave upward, this use of the doubling dose in Equation 1 will tend to overestimate risk if the data from which doubling dose is estimated are obtained from high doses. The doubling dose has traditionally been estimated from experimental animal data, mostly the mouse, although an estimate is also provided by the extensive studies of the children of atomic bomb survivors from Hiroshima and Nagasaki.

In summary, the Beir V report (NRC, 1990) states

"Although the doubling dose method is based on equilibrium considerations, the method can be used to estimate the effects of an increase in the mutation rate on the first few generations by taking a proportion of the equilibrium damage. For example, for a permanent increase in the mutation rate the effect of a dominant mutation in the nth generation is $1 - (1 - s)^n$ of the equilibrium damage, where $(1 - s)$ is the fitness of carriers of the dominant gene."

An alternative method of estimating genetic risk in the first generation is provided by the so-called "direct method" pioneered by Ehling and Selby (see Ehling, 1991). A detailed description of this method is given in UNSCEAR 1993, Appendix G. Briefly, the method is based on the equation:

$$\text{Risk per unit dose} = F_d \times M \times N \, ,$$

where:

F_d is the frequency of radiation-induced dominant mutations per unit dose,

M is the reciprocal of the fraction of total mutations thought to affect the body system(s) under study, e.g., skeletal, cataracts, and

N is the number of children born in the population under consideration.

For example, the dominant cataract mutation frequency in the mouse was estimated to be $0.15–0.18 \times 10^{-6}$ mutations per 0.01 Gy per gamete for low dose rate data. It was also estimated that approximately 2.7% of all serious dominant mutations are cataract causing mutations, i.e., M = 36.8. This gives a risk of 6–7 serious dominant disorders per 0.01 Gy of paternal exposure per 10^6 offspring. The estimate based on skeletal mutations in the mouse is similar.

Returning to the discussion of the doubling dose method, the spontaneous burden S is estimated from human epidemiologic data. The mutation component, MC, is roughly that portion of the spontaneous burden expected to increase in proportion to the mutation rate (Crow and Denniston, 1981). Dose as used in Equation 1 usually refers to the average or common dose to the gonads of both sexes, unless a sex-specific effect is being estimated.

As an example, the BEIR V committee (1990) estimated the induced burden of congenital abnormalities caused by radiation to be, after a new equilibrium is attained, 10 to 100 additional cases per million liveborn offspring per 10 mSv (1 rem) per generation (NRC, 1990). The calculations were as follows:

S = 20,000–30,000 spontaneous cases of congenital abnormalities per million liveborn offspring.

DD = 1 Sv (100 rem) for low dose or low dose rate estimated from a consideration of data from studies in mice and humans.

MC = 0.05–0.35, at the new equilibrium.

D = 0.01 Sv (1 rem) to each of the parents.

Therefore, $I = (20,000–30,000 \text{ cases}) \times (1/1 \text{ Sv}) \times (0.05 - 0.35) = 10–105$ per million liveborn, at the new equilibrium. In the report this estimate was rounded to 10–100 per million liveborn, to avoid the appearance of false accuracy. As a worst case, it was assumed that as much as 10% of this effect might manifest itself in the first generation after the increase in exposure.

To estimate the effect of increased radiation exposure on the children of exposed parents, then, one must have estimates of the spontaneous burden of the endpoint of interest, its doubling dose, the dose itself, the mutation component of the endpoint, and finally, how much of the total effect is expected to appear in the first generation after exposure.

DATA FROM WHICH RISK ESTIMATES
HAVE BEEN MADE

Mice

Studies with the mouse have yielded two kinds of results: (1) a general qualitative and semiquantitative understanding of the nature of genetic radiation effects and (2) quantitative estimates of the doubling dose. Both have been summarized in detail by the National Research Council's BEIR V committee (NRC, 1990).

The qualitative conclusions were as follows:

1. Radiation-induced mutation rates are higher in the mouse than in the fruit fly (this original finding stimulated much of the subsequent emphasis on mice because of its obvious greater relevance to estimating radiation risks in humans).

2. For mutations of specific loci (a locus is a point on a gene) induced in the spermatogonial stage, there is no significant change in the mutation rate with time after irradiation (i.e., the risk does not decrease with time after exposure).

3. Radiation-induced mutation rates differ markedly from gene to gene.

4. Mutations induced in spermatogonial and post-spermatogonial stages differ with respect to absolute and relative frequencies among loci and by radiation quality.

5. A significant proportion of the mutations detected in the specific locus test have proved to be recessive lethals.

6. Some of the recessive lethal mutations have had a heterozygote effect dramatic enough to be identified in specific individuals.

7. Dominant effects on viability are demonstrable in the first-generation progeny of irradiated males.

8. Chronic irradiation is considerably less effective than acute radiation in inducing mutations in both spermatogonia and oocytes. This dose rate effect appears to be less in males than in females.

9. A significant proportion of radiation-induced mutations in the specific locus test are small deletions.

10. The immature mouse oocyte is highly sensitive to cell killing.

Extensive literature on the mouse provides multiple endpoints from which to estimate genetic doubling doses. A detailed summary of the data can be found in Chapter 2 of the BEIR V report (NRC, 1990). The question of estimating doubling dose is discussed in the next section which includes a summary table of doubling doses for mice.

The mouse is the only mammal for which substantial data on the mutagenic effects of ionizing radiation are available. These effects have been shown to depend on dose, dose rate, fractionation pattern, LET, cell stage, sex, age at exposure, and the test stock and gene loci used. Qualitatively, these conclusions probably apply to humans as well, but whether the specific quantitative relations observed in mice transfer to humans is much less certain.

Humans

Two sets of human data have played the predominant role in estimating the risks of ionizing radiation to human populations: (1) data pertaining to the estimation of the spontaneous burden, S, in Equation 1 (Stevenson, 1959; Trimble and Doughty, 1974; Jacobs, 1975; Carr and Gedeon, 1977; Carter, 1977; Hook and Hamerton, 1977; Childs, 1981; Czeizel and Sankaranarayanan, 1984; Baird et al., 1988) and (2) data on the Hiroshima-Nagasaki atomic bomb survivors and their children used to estimate the doubling dose (Neel et al., 1953, 1974, 1990; Neel and Schull, 1956b, 1991; ABCC, 1975; Schull et al., 1981a, b; Neel and Lewis, 1990). A useful compendium of the major articles on the Japanese studies was provided by Neel and Schull (1991). Summaries and discussions of these data may also be found in reports by Denniston (1982) and UNSCEAR (1986 and 1993) and in the BEIR V report (NRC, 1990).

Estimating the Spontaneous Burden

The doubling dose approach uses the existing "normal" incidence of genetic disease as a yardstick against which to measure the effect of radiation. To do this one must know the approximate natural incidence of the endpoints under study. For example, in a sample of random births, approximately 3 percent are expected to have some kind of major congenital abnormality. Presumably, exposing the parents to additional radiation will produce additional cases over and above this spontaneous incidence.

The major studies that have provided estimates of the spontaneous burden for a number of genetic categories are provided elsewhere (Stevenson, 1959; Trimble and Doughty, 1974; Jacobs, 1975; Carr and Gedeon, 1977; Carter, 1977; Hook and Hamerton, 1977; Childs, 1981; Czeizel and Sankaranarayanan, 1984). A summary of findings is provided in Table 1 (Table 2-5 of BEIR V [NRC, 1990]).

TABLE 1. Estimated Spontaneous Burden (per 1,000 live births)

Source	Dominant	Dominant/X-Linked Combined	X-Linked	Recessive	Chromosomal Abnormality	Congenital Abnormality	Congenital Abnormalities/Multifactorial Traits Combined	Other Multifactorial Traits
Stevenson (1959)	30.7		0.2	1.0	—	10.1		10.3
UNSCEAR (1986)	9.5		0.4	2.1	4.2	25.0		15.0
BEIR (1972)	10.0		0.4	1.5	5.0	15.0		25.0
Trimble and Doughty (1974)	0.8		0.4	1.1	2.0	42.8		47.3
UNSCEAR (1977)		10.0		1.0	4.0		90.0	
Carter (1977)[a]	7.0		0.4	2.5	6.0	24.4		—
BEIR (1980)		10.0	1.1	6.0		—	90.0	—
Childs (1981)	5.8		0.3	—		—		—
UNSCEAR (1982)		10.0		2.5	6.3	43.0		
Czeizel and Sankaranarayanan (1984)				—		59.7		—
UNSCEAR (1986)		10.0		2.5	6.3	60.0		600
BEIR V (NRC, 1990)	10.0[b]		0.4	2.5	4.4[c]	20–30		1,200[d]

[a] Chromosomal abnormalities from Evans (1977).

[b] Divided into 2.5 clinically severe and 7.5 clinically mild.

[c] Divided into 0.6 unbalanced translocations and 3.8 trisomies (includes sex chromosome trisomies).

[d] Includes heart disease, cancer, and other selected disorders. The genetic component in many of these traits is unknown. To the extent that genetic influences are important, the effects are through genes that have small individual effects but that act cumulatively among themselves and in combination with environmental factors to increase susceptibility.

Again, it is important to stress that a host of genetic defects and heritable disorders will appear in any population in each generation whether or not the parents have been exposed to ionizing radiation. Radiation will tend to increase this number, but as will be seen below, the increased incidence ascribable to exposure to ionizing radiation is likely to be a very small proportion of the naturally occurring incidence.

Estimating the Effect of Ionizing Radiation

The cohort of atomic bomb survivors and their children from Hiroshima and Nagasaki is the main body of humans capable of providing estimates of the effects of ionizing radiation on the incidence of genetic disorders.

In November 1946, a presidential directive was issued at the request of the Secretary of the Navy, James T. Forrestal, giving authority to establish a Committee on Atomic Bomb Casualties. The committee was formed in January 1947 (ABCC, 1975) and was the forerunner of the Atomic Bomb Casualty Commission (ABCC), which was later transformed into the Radiation Effects Research Foundation (RERF). Its mission was "to undertake long range investigations of the effects on survivors of the bombs in Hiroshima and Nagasaki" (NRC, 1947).

The data structures and experimental designs used since the initiation of the genetic program of ABCC and RERF are described by Neel and Schull (1956a, b), Kato et al. (1966), Schull et al. (1981a), and Awa (1987). The studies were divided into four substudies:

1. The clinical program, 1948–1954. This was a prospective study of the children of atomic bomb survivors and controls involving both questionnaires and physical examinations. Five endpoints were measured: sex ratio, congenital abnormalities, viability at birth, birth weight, and survival during the neonatal period. About 92% of the children were examined as neonates and 30% were reexamined at about 9 months of age. In addition, some 717 infants who were stillborn or died in the neonatal period were autopsied. The sample included 69,706 births, of which 12,401 were from parents who were proximally exposed (i.e., were within 2,000 m of the hypocenter at the time of the bombing [ATB]).

2. F_1 mortality cohort, 1946–1985. In 1959, to increase the efficiency of the survival study, three cohorts were created from among the children born in the two cities since the bombings. The first cohort comprised all children born in the city where one or both of the parents were less than 2,000 m from the hypocenter ATB (proximally exposed). The second cohort comprised age-, sex-, and city-matched control births to parents who were more than 2,500 m from the hypocenter ATB (distally exposed). The third cohort comprised age-, sex-, and city-matched control births to parents who were not in the bombed cities ATB (not in city [NIC]). The proximal cohort contained 31,150 children, and the distal and the NIC groups numbered 41,066 children. These cohorts have been

followed through the years and form the basis not only of the F_1 mortality study but the following studies as well.

3. Cytogenetic study, 1968–present. In a subset of the F_1 cohort samples, X-chromosome anomalies and balanced structural rearrangements were looked for in the blood of the children of proximally exposed, distally exposed, and NIC parents. All children were at least 13 years of age when the samples were obtained.

4. Biochemical studies, 1975–1992. In a subset of the F_1 mortality cohort, a direct search was made for new mutations by using a battery of 30 serum and erythrocyte proteins.

Overall, eight health outcomes have been investigated:

1. Untoward pregnancy outcomes: congenital malformations, infant stillbirth or death within the first 2 weeks after birth.

2. F_1 mortality: death in children of exposed parents after 2 weeks, exclusive of cancer.

3. Malignancies in the F_1 cohort: cancer arising in the children of survivors. Some cancers are the result of a combination of germinal and somatic mutations. Mutations induced by radiation might be detected by observing an increase in the incidence of such cancers.

4. Balanced structural rearrangements of chromosomes in children over age 13 years: because the children from whom samples were obtained had all reached at least the age of 13 years, only balanced rearrangements would be expected.

5. Sex chromosome aneuploids in children over age 13 years: Individuals with the sex chromosome anomalies XXY, XYY, XO, and XXX are all viable, although some would not be expected to survive to age 13 years.

6. Mutations altering protein charge or function: this program centered on the detection of rare protein variants, in which case studies involving the family were carried out to determine whether the variant had been inherited or was the result of a mutation in the preceding generation. Collectively, 1,256,555 locus tests were done, and among these, seven apparent mutations were detected. Four of these occurred among the children of exposed parents and three occurred among the children of the controls.

7. Sex ratio in children of survivors: the proportion of male births among parents exposed to different amounts of ionizing radiation.

8. Growth and development of children of survivors: birth weight and weight, body length, head circumference, and chest circumference at 8–10 months of age.

Overall, the studies of health outcomes in the Hiroshima-Nagasaki atomic bomb survivors and their children have revealed a small but statistically nonsignificant difference in health outcomes between the children of the atomic bomb survivors conceived subsequent to the bombing and the children of nonexposed parents. This increase, albeit small, is qualitatively and quantitatively consistent with the known mutagenicity of ionizing radiation in experimental animals and provides the best currently available basis for estimating the doubling dose from human data (Neel and Schull, 1991).

ESTIMATING THE DOUBLING DOSE

To make use of Equation 1, it is necessary to estimate a genetic doubling dose. This has been done in a number of original studies and also by the various committees assigned the task of evaluating the genetic risks of radiation. The idea is straightforward. If one assumes that the relation between mutation rate and dose is linear, at least at low doses the model may be written as

$$M = m_1 + m_2 + b_1D_1 + b_2D_2, \text{(Equation 2)}$$

where 1 refers to males and 2 refers to females. Here m_1 and m_2 are the spontaneous mutation rates in males and females, b_1 and b_2 are the induction rates in males and females, and D_1 and D_2 are the doses applied to the two sexes. If applied to both sexes, the common dose would be $M = 2(m_1 + m_2)$, therefore, the doubling dose (D), is $(m_1 + m_2)/(b_1 + b_2)$. This is estimated by the regressing effect on the average dose to the two sexes, $M = \alpha + \beta D$, and obtaining the doubling dose from the estimation equation $DD = \alpha / \beta$, where α is the intercept and β is regression.

Mice

Table 2 (Table 2-11 of BEIR V [NRC, 1990]) contains estimates of the doubling doses for chronic (low dose rate) ionizing radiation for a number of different endpoints. The ranges in parentheses were obtained by the BEIR V committee by multiplying acute doubling dose estimates by a correction factor range of 5–10. The figures in this table are based on a large number of studies of different sizes and reliabilities. The reader should refer to the original studies (references given in BEIR V) before making use of individual estimates.

The overall median estimate is in the range of 1.0–1.14 Sv (100–114 rem) for chronic exposure. The median acute doubling dose estimate is about 0.30 Sv (30 rem).

Humans

The Japanese data have been used to estimate minimum and probable genetic doubling doses in humans (Neel et al., 1974, 1990; Neel and Lewis, 1990).

Table 3, modified from Table 5 of Neel et al. (1990), contains the most recent estimates of minimum acute doubling dose on the basis of data from Hiroshima-Nagasaki atomic bomb survivors. The numbers in the last three columns are the lower 99, 95, and 90 percent confidence limits of the doubling dose for five endpoints: untoward pregnancy outcome (UPO), F_1 mortality, F_1 cancer, sex chromosome aneuploids, and loci-encoding proteins. These lower 95 percent confidence limits range from 50 mSv (5 rem) to 2,270 mSv (227 rem). In addition, Neel et al. (1990) suggest a range of point estimates for acute doubling doses of from 1,690 mSv (169 rem) to 2,230 mSv (223 rem) and for chronic (low-dose-rate) ionizing radiation exposure of 3,380 mSv (338 rem) and 4,660 mSv (466 rem).

TABLE 2. Estimated Doubling Doses for Chronic Radiation Exposure (primarily mouse)

Genetic Endpoint and Sex	Doubling Dose (rads)[a]
Dominant lethal mutations,	
Both sexes	40–100
Recessive lethal mutations,	
Both sexes	(150–300)
Dominant visible mutations	
Male	
Skeletal	(75–100)
Cataract	(200–400)
Other	80
Female	(40–160)
Recessive visible mutations	
Postgonial, male	
Postgonial, female	70–600
Gonial, male	114
Reciprocal translocations	
Male	
Mouse	10–50
Rhesus monkey	(20–40)
Heritable translocations	
Male	(12–250)
Female	(50–100)
Congenital malformations	
Female, postgonial	(25–250)
Male, postgonial	(125–1,250)
Male, gonial	(80–2,500)
Aneuploidy (hyperhaploids)	
Female	
Preovulatory oocyte	(15–250)
Less mature oocyte	(250–1,300)
Median (mouse, all endpoints, both s	
Direct estimates	70–80
Indirect estimates	(150)
Overall	100–114

[a] Values not in parentheses are based on the spontaneous rate divided by the induced rate/rads for the low dose rate; values in parentheses are based on the spontaneous rate divided by the induced rate/rad at the high dose rate, multiplied by a factor of 5–10 to correct for the dose rate effect.

It is important to note that these doubling dose estimates for humans and their lower limits refer to "conjoint" doubling doses, that is, the sum of the parental doses that is expected to double the genetic burden. The doubling dose estimate for the mouse given in Table 2 and the doubling dose calculated from Equation 1, as customarily used, refer to the common or average dose exposure to each of the two parents that is expected to double the genetic burden. For example, a common exposure of 10 mSv (1 rem) to each of the two parents corresponds to a conjoint dose of 20 mSv (2 rem). Consequently, to compare the doubling doses for mice with the lower bounds and point estimates of doubling doses for humans, either divide the figures for human by 2 or multiply the figures for mouse by 2. Both are perfectly valid doubling doses; they are simply scaled differently. Either can be used in Equation 1, so long as DD and D are used consistently.

It appears that humans may be less sensitive to the mutagenic effects of ionizing radiation than mice. Neel and Lewis (1990) have recently attempted to resolve this difference by suggesting that, overall, the mouse estimates of doubling doses are actually higher than those suggested in Table 2.

In sum, the general scientific consensus is that the overall doubling dose of mutation induction for low-LET, low-dose ionizing radiation is on the order of 100 rem, and it may, in fact, be larger.

Table 3. Estimate of Doubling Doses That Can Be Excluded at Specified Confidence Levels

Trait	Observed Total Background	Mutational Component (%)	Regression (β)	Intercept (α)	Doubling Dose (Sv) at Lower Confidence Limit of:		
					99%	95%	90%
UPO[a]	0.0502	3.4–5.4	.00264 ± .00277	.03856 ± .00582	0.14–0.23	0.18–0.29	0.21–0.33
F1 mortality	0.0458	3.5–5.7	.00076 ± .00154	.06346 ± .00181	0.51–0.83	0.68–1.10	0.81–1.32
F1 cancer	0.0012	2.0–4.0	−.00008 ± .00028	.00104 ± .00033	0.04–0.07	0.05–0.11	0.07–0.15
Sex chromosome aneuploids	0.0030[b]	100	.00044 ± .00069	.00252 ± .00043	1.23	1.60	1.91
Loci encoding proteins	0.000013	100	−.00001 ± .00001	.00001 ± .00001	0.99	2.27	7.41

[a] UPO, Untoward Pregnancy Outcome: congenital malformations, stillbirths or deaths under 2 weeks of age.
[b] Observed Zygotic Mutation Rate.

7

Adverse Reproductive Outcomes

The term "adverse reproductive outcome" includes such diverse endpoints as the inability to conceive (sterility or infertility), the premature spontaneous termination of a pregnancy (abortion), the birth of an infant with a congenital malformation or with mental or physical retardation, and the premature death of an offspring (stillbirth, neonatal, or infant death). Many host and environmental factors contribute to the origins of such outcomes. They may be caused by maternally or paternally derived inherited defects, exposure to noxious environmental agents, including ionizing radiation, smoking, or the consumption of alcohol, preexisting maternal illness (such as diabetes) or illness during pregnancy, and malnutrition (Bracken, 1984). This diversity of possible origins makes difficult the assignment of causation in any specific instance.

Of primary interest here are those adverse reproductive outcomes that may have arisen through the induction of a deleterious mutation in the paternal germ cells as a consequence of exposure to ionizing radiation. As noted in Chapter 6, this contribution to the totality of adverse reproductive outcomes is called the mutation component, and it varies substantially from one endpoint to another. Unfortunately, in most instances the precise size of this mutation component is either unknown or poorly estimated, but sufficient information is available to estimate its probable magnitude.

GENERAL REMARKS

In the paragraphs that follow, the committee examines separately each of 11 adverse reproductive outcomes. This examination will include a definition of the endpoint, current estimates of its frequency, its cause(s), possible sources of extraneous variability (confounders) that may contribute to its prevalence or incidence, the interaction of various contributors to the occurrence of the endpoint, and the difficulties inherent in obtaining reliable estimates of prevalence or incidence and paternal exposure. The risk factors described primarily relate to the mother's preconceptional or gestational exposure. Although much of what is known about the causes of these endpoints relates to maternal exposure, specific reference will be made to the status of evidence for male-mediated effects. In addition, these maternal risk factors are discussed because they are potentially strong confounders that would have to be taken into account in any study of paternal radiation exposure. A confounder in this context is a variable that is causally related to the disease under study and that is associated with exposure in the study population, but that is not a consequence of this exposure (Kelsey et al., 1986).

Current research on adverse reproductive outcomes has been expanded to include effects on the reproductive systems of both men and women as well as effects on the offspring. This expanded list of effects is shown in Table 4 and includes menstrual cycle changes; semen characteristics; fecundability and fertility; embryonic and fetal loss; any complication affecting the embryo, fetus, or mother; infant morbidity and mortality; and childhood malignancies. For example, preconceptional exposure may lead to spermatic or ovarian toxicity and germ cell mutations in either sex. These latter effects may in turn lead to infertility, spontaneous abortions, or abnormalities in the offspring. Prenatal exposures could lead to functional deficiencies and illnesses in the offspring that may not become apparent until childhood or even adulthood, such as developmental disorders and malignancies. In summary, the reproductive and developmental effects of environmental agents may operate through a variety of mechanisms including toxic, mutagenic, teratogenic, and carcinogenic effects.

TABLE 4. Possible Reproductive Effects of Environmental Exposures

1. Menstrual cycle disorders	9. Intrauterine growth retardation
2. Hormonal changes	10. Preterm birth
3. Sperm or semen abnormalities	11. Maternal complications
4. Infertility	12. Birth defects
5. Single gene defects	13. Neonatal and infant deaths
6. Chromosomal aberrations	14. Developmental deficits
7. Fetal loss	15. Childhood malignancies
8. Altered sex ratio	16. Other childhood diseases

SOURCE: Adapted from Berkowitz (1985).

This multiplicity of possible adverse reproductive effects and etiologies makes difficult the design and implementation of epidemiologic studies seeking to establish the association of a specific effect(s) with a given exposure, as well as the meaningful interpretation of a suspected association.

MALE-MEDIATED REPRODUCTIVE AND
DEVELOPMENTAL OUTCOMES

The feasibility of studying adverse reproductive outcomes among Atomic Veterans and their families hinges in large measure on the likelihood that genetic effects may be seen following exposure of the male to ionizing radiation. To assess this likelihood the committee summarized the evidence on paternal exposures and adverse reproduction and disease in offspring found in studies of animals and humans.

The role of paternal exposure in the origin of many adverse reproductive and developmental outcomes has not been investigated extensively in humans (Olshan and Faustman, 1993). The effects of chemicals or ionizing radiation on the sperm, chromosome, and fertility have been demonstrated (Wyrobek et al., 1983; Martin et al., 1986; Geneseca et al., 1990). The prevailing view is that exposure of the human male to chemicals and ionizing radiation is generally unrelated to the occurrence of developmental endpoints such as miscarriage, birth defects, growth retardation, and cancer (Brown, 1985). Animal studies, some repeatedly confirmed, demonstrate that paternal exposure can lead to a variety of effects, including skeletal malformations and cataracts, in the offspring. A causal connection has been established with radiation-induced mutations. There is a paucity of human data on this subject.

Several potential mechanisms have been proposed to explain possible male-mediated effects on offspring. A direct effect of an agent on male germ cell DNA is the traditional explanation for the induction of some abnormalities and is of major relevance for considering the potential effects of ionizing radiation. Since the time span for male germ cells to mature into spermatozoa is only about 8–9 weeks (compared to the longevity of a lifetime for resting oocytes), the effect of environmental agents must be on the spermatogonia for possible effects to be long lasting. The majority of the available data from tests in animals suggest that this mechanism is involved.

More indirect mechanisms involving the transfer of toxic agents in seminal fluid and maternal exposure to agents brought home by the father have been suggested. Some toxicants, such as PCBs and lead, have been found in seminal fluid, and vaginal absorption of substances in semen can occur. Nevertheless, it is unclear whether the concentrations of these toxicants obtained through this mechanism would be sufficiently great to have an effect on the fetus (Hatch and Marcus, 1991). There are some data from studies in animals linking seminal transfer of chemicals to preimplantation loss (Robaire and Hales, 1994). There is no evidence of seminal fluid toxicants being produced as a result of radiation exposure.

As noted above, evidence for the direct effects of exposure on male germ cell DNA (germ cell mutagenicity assays) from studies in experimental animals has been available for several decades. In fact, an important test in animals for germ cell mutagenicity, the specific locus test, was developed in 1951 and has been extensively used in tests involving ionizing radiation (Russell, 1951). The majority of the data from tests in animals were obtained in the evaluations of radiation and chemicals in relation to the expression of defined visible phenotypes caused by mutations at recessive loci (specific locus test), fetal loss (dominant lethal test), inherited chromosomal aberrations (heritable translocation test), and congenital anomalies (dominant skeletal and dominant cataract tests). Ionizing radiation and a small number of chemicals (compared to animal carcinogenicity assays) have yielded positive results in these test systems. Data from tests in animals also show that exposure of males to ionizing radiation and some chemicals can produce other outcomes in the offspring such as tumors, growth retardation, and neurobehavioral effects (Olshan and Faustman, 1993). Detailed data from studies in experimental animals exposed to ionizing radiation can be found in Chapter 6.

Epidemiologic associations between paternal exposure and reproductive and developmental abnormalities in offspring have been reported, although further verification is needed. Most of the studies have focused on occupational exposures. A brief summary of the findings from epidemiologic studies is provided here. The major epidemiologic studies relating to ionizing radiation and developmental outcomes, such as the studies of the atomic bomb survivors, can be found in Chapter 6. There is evidence that a small number of agents such as ionizing radiation and dibromochloropropane (DBCP) may affect semen quality and result in reduced fertility or sterility when individuals are exposed to high doses. An increased risk of fetal loss (spontaneous abortion) has been linked with paternal occupational exposure to vinyl chloride, anesthetic gases, dibromochloropropane, mercury, lead, other metals, and various solvents (Savitz et al., 1994). A number of epidemiologic studies have examined the relationship between paternal occupation and birth defects (Olshan and Schnitzer, 1994). Positive associations have been found for individuals with a variety of occupations, including painters, welders, firemen, forestry and logging workers, motor vehicle operators, wood workers, farm workers, and metal workers. Individuals in these occupations have a variety of exposures, often mixed, to metals, solvents, pesticides, and paints.

Several recent large case–control studies have reported some paternal occupations and exposures that may be associated with childhood cancer in the offspring (Savitz, 1986; Savitz and Chen, 1990; O'Leary et al., 1991). These associations include leukemia in the offspring of painters, mechanics, and machinists; childhood brain tumors in the offspring of painters, metal workers, those in electronics-related occupations, and those with motor vehicle-related jobs; Wilms' tumor (childhood kidney tumor) in the offspring of auto mechanics and machinists, welders, and painters; and neuroblastoma in the offspring of those in electronics-related occupations.

To summarize briefly, some male-mediated environmental exposures may affect pregnancy outcome through fertilization by mutation-bearing sperm, which could lead to pregnancy loss or possibly congenital malformations or childhood cancer (Berkowitz and Marcus, 1993). It is also possible that male-mediated effects could occur because male workers bring toxicants into the home environment or because toxicants are absorbed into the female genital tract via sperm or seminal fluid. The experimental animal and epidemiologic data indicate that the exposure of the male to various toxic agents may increase the risk of the full spectrum of adverse developmental endpoints from fetal loss to cancer. However, the evidence is not firm and clearly requires more study and confirmation in both laboratory and epidemiologic settings.

INFERTILITY

Approximately 15% of all couples of reproductive age cannot achieve a desired pregnancy (McClure, 1986). The large majority of previous research has concentrated on the female component; however, half of all fertility problems may be due to male reproductive dysfunction (Swerdloff, 1985). The role of extrinsic environmental factors on male reproduction has suggested that chemical and physical exposures have an effect (Biava et al., 1978; Steeno and Pangkahila, 1984; Bonde, 1990). In a recent study, occupationally related exposure to electromagnetic fields was unrelated to morphology, motility, or concentration of semen (Lundsberg et al., 1995). Ionizing radiation at high enough doses can lead to temporary or permanent sterility. The doses and time course of the process in humans have been studied extensively. In general, the human data related to testicular effects are reported from accidental exposures or from males irradiated for therapeutic reasons, for example, as a treatment for testicular cancer. Following the administration of doses as low as 80 mGy (8 rads), a reduction in the number of spermatogonia occurs. At these low doses, the more resistant and more differentiated cells in the line may continue normal maturation. A reduction in the sperm count may not be evident until 30 to 60 days have passed. The ultimate degree and duration of depletion of the sperm depend on the magnitude of the dose received.

Sterility following irradiation is often a loosely applied term; it may represent complete sterility, temporary subfertility, or in fact, fertility with a reduced number of sperm. Investigation of human volunteers indicates that administration of 25 fractions of 150–200 mGy (15–20 rad) daily may cause a decrease in the sperm count. It is of interest that in the few patients treated with radiotherapy to one testicle or to the inguinal nodes at doses of 600–2,500 mGy (60–250 rad), a significant percentage were subsequently able to father children. Lushbaugh and Casarett (1976) have reviewed the literature and have found no case reports of malformed infants from parents who had received radiotherapy before the conception of the child.

Five to six Gy (500 to 600 rad) in a given dose will cause permanent sterility in most men; however, in some cases, that dose has been exceeded without causing permanent sterility. A dose of 2.5 Gy (250 rad) may cause transient sterility for about 12 months. UNSCEAR (1982) has reviewed the available literature and concludes that

(1) temporary sterility is reported to occur with single doses to the testes ranging from 1.5 to 4 Gy (150 to 400 rad) and with fractionated doses of 0.1–2 Gy (10 to 200 rad), and (2) permanent sterility occurs with single doses ranging from 5 to 9.5 Gy (500 to 950 rad) and with fractionated doses of 2–6 Gy (200–600 rad).

A person may have transient sterility for a matter of years before fertility returns. Five individuals who received doses of 2.3–3.7 Gy (230–370 rad) in the Oak Ridge criticality accident were aspermic for 4 months and hypospermic for 21 months; at least one individual experienced "sterility" for several years, but the individual subsequently had a normal offspring (Andrews et al., 1980).

Ortin et al. (1990) have reviewed data on 148 boys treated for Hodgkin's disease and followed for a median of 9 years. Sexual maturation was achieved in all boys without the need for androgen replacement. Of eight boys who were treated with radiation alone, three who received a pelvic dose of 40–45 Gy (4,000–4,500 rad) were able to father children. Three others who received 30–44 Gy (3,000–4,400 rad) of pelvic radiation were oligospermic. This was in contrast to an 83% incidence of absolute azospermia in boys treated with chemotherapy and no pelvic radiation. Patients who have received whole body irradiation before bone marrow transplantation have been studied by Sanders et al. (1986) and Deeg et al. (1984), who report that gonadal failure occurred in almost all boys who were postpubertal at the time of irradiation.

Radionuclide Irradiation

Irradiation of the testes by radionuclides may result from internal or external exposure. In the case of energetic beta radiation, the germ cells may be irradiated by external radionuclides. Although there may be internal deposition of radionuclides such as cesium, few data on the human sperm count or sterility are available. Although some radionuclides are known to be preferentially deposited in specific organs or tissues, such as radioiodine in the thyroid or strontium in bone, no radionuclide is known that is preferentially deposited in the testes.

SPONTANEOUS ABORTIONS

The World Health Organization (1970) has defined spontaneous abortion as "the non-deliberate interruption of intra-uterine pregnancy before 28 weeks (LMP) in which the embryo or fetus is dead when delivered." However, in many technically developed countries 22 weeks of gestational age separates spontaneous abortion from stillbirth (Bracken, 1984).

Spontaneous abortions occur much more frequently in early pregnancy, with most thought to occur before pregnancy is recognized. Spontaneous abortion rates as high as 80% of all conceptions have been suggested, but the most widely cited rates are 30–35% (Wilcox et al., 1988). Of clinically recognized pregnancies, approximately 15% will spontaneously abort. Second-trimester spontaneous abortion occurs in 1–3% of all pregnancies.

There is considerable heterogeneity of spontaneous abortion that varies by the gestational age of pregnancy. As many as 95% of very early abortions are of congenitally malformed embryos and a large proportion of these have chromosomal abnormalities. Of the clinically recognized pregnancies that spontaneously abort, 65% are chromosomally abnormal and include fetuses with developmental anomalies.

The risk of spontaneous abortion is higher with increased gravidity, and this may relate to such maternal factors as retroversion of the uterus, fibroids, prolapsed uterus, cervical erosion, and uterine incompetence. A septate or bicornate uterus will also increase the risk of spontaneous abortion. There is a reoccurrence rate of about 1.6% which is found for abortions of both euploid and aneuploid fetuses. Repeat spontaneous abortion (sometimes called habitual abortion) has also been shown to relate to both very early or very late age of first menstrual cycle.

Other known or suspected maternal risk factors for spontaneous abortion are maternal smoking, which primarily increases the risk of abortion of euploid fetuses; maternal alcohol use; occupational exposure to lead, vinyl chloride, and solvents; exposure to antineoplastic drugs; cocaine use; and heavy maternal caffeine use. Maternal infection with malaria, rubella, rubeola, herpes virus, cytomegalovirus, and genital mycoplasmas has been associated with an increased risk of spontaneous abortion. Recent use of oral contraceptives or a diaphragm has been related to a decreased spontaneous abortion risk. The frequency of spontaneous abortion was studied in atomic bomb survivors, but the data were internally inconsistent (Neel and Schull, 1991).

PRETERM DELIVERY

Preterm delivery is a major cause of neonatal morbidity and mortality. Traditionally, prematurity was defined as birth weight of less than or equal to 2,500 g. Today, a distinction is made between low birth weight births and preterm births. Low birth weight characterizes an infant who weighs less than 2,500 g (5 pounds 8 ounces) at birth, and preterm refers to a birth that occurs at a gestational age of less than 37 completed weeks (<259 days). The rate of preterm delivery in the United States is approximately 10% (Berkowitz and Papiernik, 1993). The rate has varied only slightly between 9 and 10% since 1970.

Established maternal risk factors for preterm birth include African-American race, single marital status, low socioeconomic status (SES), previous low birth weight or preterm delivery, multiple second-trimester abortions, cigarette smoking, in vitro fertilization pregnancy, and such gynecologic or obstetrical complications as in utero diethylstilbestrol exposure, cervical and uterine anomalies, gestational bleeding, and placental abnormalities. In addition, urogenital infections, cocaine use, and no or inadequate prenatal care are probably associated with preterm delivery. Other factors such as age, parity, and maternal weight gain appear to be weakly associated or not associated with preterm delivery.

Still, the association remains inconclusive for other factors such as low prepregnancy weight, physical activity, and psychological stress (Berkowitz and Papiernik, 1993). The frequency of congenital malformations is higher in preterm than in full-term pregnancies (Placek, 1977; Hartikainen-Sorri and Sorri, 1989). Standing for long periods of time at work (Teitelman et al., 1990) and a history of asthma and bronchodilator use during pregnancy have also been implicated as risk factors for preterm delivery (Doucette and Bracken, 1993). A short interval between pregnancies has recently been reported to be a possible risk factor for delivery of a preterm, low birth weight infant and, potentially, as an important explanatory factor for the racial disparity in adverse pregnancy outcomes (Rawlings et al., 1995).

Neither paternal nor maternal exposure to low dose radiation is considered to be a risk factor for preterm birth. There are reports that the number of preterm births in the seventh to eighth month were increased in contaminated areas shortly after the Chernobyl nuclear power plant accident in 1986 (IAEA, 1991). However, it is not clear to what extent this may have been due to such factors as maternal nutrition, stress, cigarette smoking, and alcohol consumption. No effect of parental irradiation on mean birth weight could be demonstrated in survivors of the atomic bombs at Hiroshima and Nagasaki (Neel and Schull, 1991).

STILLBIRTHS AND NEONATAL DEATHS (PERINATAL DEATHS)

Stillbirth is widely accepted to be a fetal death occurring at 28 weeks of gestation or later (from the last menstrual period [LMP]). Because both stillbirth and early neonatal death (under 7 days of age) share in a number of causes, which are themselves somewhat different from causes of death in older neonates, the term "perinatal death" is often used to refer to both of them. In 1977, the World Health Organization recommended including all infants weighing at least 500 g or, if the birth weight is unavailable, infants delivered at 22 or more weeks of gestation as stillbirths. Occasionally, all deaths of infants delivered at 16 weeks or more of gestation are included as stillbirths.

Perinatal (neonatal) mortality rate is typically measured as the number of deaths occurring within the first 28 days of life per 100,000 live births. Neonatal mortality has declined substantially in the past 50 years. In the United States in 1992, neonatal death rates were 537.5 per 100,000 live births; however this differed significantly among black and white infants (rates of 1,083.1 per 100,000 live births and 434.6 per 100,000 live births, respectively) (DHHS, 1995).

The higher perinatal (neonatal) mortality rate (PMR) is seen in lower socioeconomic (SES) groups, but it is difficult to disentangle the behavioral, social, and environmental factors that correlate with socioeconomic class. PMR is consistently higher in unmarried women and women who smoke, and PMR decreases with parity. PMR increases with maternal age and variety of maternal diseases; including diabetes, blood group incompatibility, uterine infections, renal disease, and preeclampsia. Pregnancy complications include placenta previa and abruptio, pelvic anomalies, malpresentation, prolonged duration of labor, uterine rupture, umbilical cord complications, and fetal complications. These maternal factors are among those that should be taken into account in evaluating paternal influences on PMR.

Results from the atomic bomb survivors show that high doses had no effects on PMR (Neel and Schull, 1991), so no effect would be expected at the much lower doses received by Atomic Veterans.

BIRTH DEFECTS

Conventionally, a birth defect (a congenital malformation) implies a failure of proper or normal morphologic development. Such failures can vary greatly in the threat that they pose to an individual's physical and mental integrity and survival. As a result, it is common to divide malformations into major and minor malformations although there is no absolute dividing line between these two classes of defects. Major birth defects include those failures of normal development that are incompatible with life (anencephaly, for example), that are seriously life-threatening (such as many congenital heart defects), or that materially compromise the individual's ability to function effectively in the society of which he or she is a member (cleft palate). All other birth defects are construed as minor.

Birth defects, whether major or minor, are often further classified in two ways: (1) malformations arising from anomalies of a gene or a chromosome and (2) developmental disorganization occurring in an embryo or fetus with a normal genotype.

Human Chromosome Abnormalities

Four subtypes of chromosomal abnormalities can be considered on the basis of whether the affected cell is somatic or germinal and whether the anomaly is structural or numeric. Many chromosomal abnormalities are incompatible with embryonic or fetal survival and are never expressed in the human phenotype at delivery, although some (e.g., trisomy 16) may be relatively common in early spontaneous abortion.

Table 5 shows the prevalence of the more frequently seen chromosomal abnormalities in individuals born to parents exposed to ionizing radiation at Hiroshima and Nagasaki and to controls. The data were tabulated by Hook (1984) and were collected by Awa (for a recent summary of the findings see Awa et al., 1987, 1988).

The offspring of exposed parents (n = 8,322) and control parents (n = 7,976) who were delivered between May 1946 and December 1984, are being studied at about age 13 years. This late age at examination screens out many of the chromosomal anomalies that might incur early mortality, and thus, the data provide reliable estimates on only two types of chromosomal abnormalities, namely, sex chromosome aneuploids and balanced structural rearrangements, which confer relatively little threat to survival among liveborn infants. The data presented do not suggest any major differences according to exposure. However, when the frequency of sex chromosome aneuploidy is examined in the context of the combined parental gonadal dose, there is a small but statistically nonsignificant increase in these anomalies with increasing parental dose (Neel et al., 1990).

Developmental Abnormalities

Developmental abnormalities can affect any organ, and as a result, there are several hundred potential diagnostic categories. Early fetal lethality reduces the number of malformations seen at birth. The most frequent developmental malformations and their prevalence in those under age 1 year are given in Table 6 (Bracken, 1983). The most commonly observed malformations at birth are of the ventricular septum of the heart; the majority of malformations occur at rates of <2 per 1,000 live births (Mellin, 1963). Overall, in some 3% of newborns a major congenital anomaly will be diagnosed at birth, and an additional 3–5% will be diagnosed with a major congenital anomaly in the first 10 years of life. If the Atomic Veterans each fathered an average of two or three children, the total would be about 500,000 offspring. The rates for birth defects given in Table 6 can be used to estimate the numbers of birth defects that would be seen among these children of Atomic Veterans *in the absence of any radiation effects.* For example, spina bifida is routinely observed in 1.4 in 1,000 babies, which would be 700 among the infants in 500,000 offspring of Atomic Veterans. Among the offspring of Atomic Veterans there would be 5,000 infants with heart defects, and nearly 600 would be diagnosed with cancer during childhood. In total, one would expect to find 15,000 infants with major malformations among the children of Atomic Veterans if the rates among those children were the same as those in the general population.

A subcategory of malformation, sometimes called deformational, are those in organ systems that developed normally but that incurred a secondary deformity. Usually, this is due to intrauterine molding because of oligohydramnios and physical constraint (examples of these are some types of talipes, congenital hip dislocation, and midline cleft palate).

TABLE 5. Chromosomal Abnormality Prevalence in Individuals Born to Parents Exposed to Radiation at Hiroshima or Nagasaki and to Controls[a]

Abnormalities	Father Only	Mother Only	Mother and Father	Either or both Parents	Controls
Sex chromosome abnormalities[b]					
Male					
XYY	3.0	0.6	0	1.1	1.3
XXY	1.5	2.5	2.4	2.3	2.6
Other	0	0.6	0	0.4	0
Female					
XXX	2.6	0.5	2.2	1.3	0.7
45,X	0	0	0	0	0
Other	0	0.5	2.2	0.6	0.7
Subtotal (all sexes)[c]	3.5	2.3	3.4	2.8	2.6
Autosomal trisomy	0	0	0	0	0
Autosomal structural rearrangement					
Balanced Robertsonian (Dq/Dq)	0.7	0.9	0	0.7	0.2

(Dq/Gq)	0	0.6	0	0.3	0
(Gq/Gq)	0	0	0	0	0
Reciprocal	0.7	0.3	2.3	0.7	1.6
Inversion	0	0.3	0	0.2	0.4
Other	0	0	0	0	0
Subtotal (autosomal structure)	1.4	2.1	2.3	1.9	2.2
Unbalanced autosomal supernumerary	0.7	0	0	0.2	0.2
Other autosomal	0	0.6	0	0.3	0
Subtotal (all autosomal)	2.1	2.7	2.3	2.4	2.4
Grand total	5.6	5.0	5.7	5.2	4.9
Number of cases					
Males	667	1,585	410	2,662	2,267
Females	775	1,861	464	3,100	2,791
Both	11,442	3,446	874	5,762	5,058

[a] Values are rates per 1,000 individuals.

[b] Rates of sex chromosome abnormalities are sex specific (e.g., rates of XYYs are in males only).

[c] Rates in subtotal are both sexes.

SOURCE: Modified from Hook (1984).

The range of risk factors studied for their effects on developmental malformations is too large to summarize in any detail here. Several maternal infections (e.g., rubella, cytomegalovirus, toxoplasma, and herpes virus) have been associated with an increased risk of malformations and may account for 2% of all malformations. Maternal diabetes is linked to a range of malformations. Numerous environmental substances have been studied (usually, inadequately) but with inconclusive results. Lead at relatively high doses is a behavioral teratogen, but it is not related to physical malformations. Mercury, especially in organic forms (methyl mercury), has some physical teratogenic properties, but it is primarily a neurobehavioral teratogen. Maternal alcohol and cocaine use are related to a range of malformations at moderate levels of exposure. Cigarettes and caffeine do not appear to be teratogenic at their usual levels of intake. Only a few pharmaceutical agents have clearly been shown to be teratogenic, including diethylstilbestrol, thalidomide, phenytoin, and some other antiepileptic drugs. Warfarin and some cytotoxic drugs have also been linked to developmental anomalies. Nutritional deficiency, particularly folate deficiencies, increases the frequency of developmental anomalies. These maternal exposures should be excluded in considering the birth defects attributable to paternal exposures. There was no evidence of an increased risk of congenital malformations following paternal exposure to ionizing radiation or in children of atomic bomb survivors (Neel and Schull, 1991).

MATERNAL ILLNESSES

The committee has interpreted periparturient diseases of the mother as disorders that can cause maternal and neonatal morbidity. Numerous chronic and pregnancy-related disorders in the mother may affect the mother's and neonate's health and well-being. These include infections such as cytomegalovirus, herpes simplex virus, and hepatitis; maternal cardiovascular disease including congenital and rheumatic heart diseases; hypertensive disorders; pulmonary disorders such as pneumonia and asthma; diabetes mellitus, including gestational diabetes; hyper- and hypothyroidism; maternal adrenal gland disorders such as Cushing syndrome and Addison disease; collagen vascular disease such as systemic lupus erythematosus; renal diseases such as glomerulonephritis and nephritis; placenta previa, abruptio placenta, and retained placenta; hematologic disease including idiopathic thromobocytopenic purpura and Rh incompatibility; myasthenia gravis; and epilepsy and other neurologic diseases. None of these diagnoses, however, are known to be related to irradiation of either parent.

Table 6. Prevalence per 1,000 Live Births of Congenital Malformations by Diagnostic Group from Five Major Hospitals in Connecticut, 1974–1976 and Expected Prevalence at Birth for a Population of 500,000 Newborns[a]

Diagnostic Group	Number per 1,000 (All Diagnosed)	Expected Number Among 500,000 Births
All malformations	57.01	28,505
All neoplasms	1.16	580
Hemangiomas and lymphangioma	2.06	1,030
Strabismus	0.87	435
Heart block fibrillation, tachycardia	0.33	165
Inguinal hernia plus obstruction	7.68	3,840
Anencephaly	0.41	205
Spina bifida	1.40	700
Hydrocephaly	1.40	700
Eye anomalies	0.54	270
Common truncus	0.21	105
Transposition of great vessels	0.99	495
Tetralogy of Fallot	0.70	350
Ventricular septal defect	6.07	3,035
Atrial septal defect	0.50	250
Heart valve	1.73	865
Other heart anomalies	0.87	435
Total heart anomalies	11.07	5,535
Patent ductus arteriosus	2.19	1,095
Coarctation of aorta	0.62	310
Single umbilical artery	0.78	390
Cleft lip palate	1.86	930
Pyloric stenosis	8.01	4,005
Tracheal-esophageal fistula	0.45	225
Other digestive system anomalies	1.69	845
Undescended testicles	0.62	310
Hypospadias	1.40	700
Congenital hydrocele	1.57	785
Other genital organ anomalies	1.03	515
All talipes	3.30	1,650
Polysyndactyly	2.19	1,095
Limb reduction	0.66	330
Congenital dislocation of hip	0.99	495
Lower limb anomalies	0.21	105
Skull and face anomalies	0.62	310
Other muscular-skeletal anomalies	1.16	580
Skin anomalies	2.62	1,310
Down syndrome	1.36	680
Other autosome anomalies	0.58	290
Other unspecified multiple anomalies	1.28	640

[a] Data are based on malformations found among 24,224 live births in five Connecticut Hospitals over a 24-month period.

SOURCE: Modified from Bracken (1983).

ALTERED SEX RATIO

When studies of the Hiroshima-Nagasaki atomic bomb survivors and their offspring began, it was believed that a person's gender was determined in a simple way. Individuals inheriting an X chromosome from their father were destined to be females, whereas those individuals who inherited a Y chromosome from their father would be males. Thus, females would have two X chromosomes and males would have only one. These notions suggested, in turn, that when mutations in genes on the X chromosome induced by ionizing radiation are incompatible with survival (are lethal), their expression would be manifested differently in the two genders. More specifically, since a father normally transmits his single X chromosome to his daughters, if a lethal mutation were present on the X chromosome in the father's sperm, it could find expression only in his daughters. On the other hand, since mothers transmit their X chromosomes equally to their sons and daughters, a lethal mutation might be expressed in either sex. If the mutation was dominant, the two sexes would be affected equally often; however, if the mutation was recessive, since the male has only one X chromosome, it would invariably manifest itself in males, but in females manifestation of the new mutant would occur only if the second X chromosome fortuitously carried a functionally similar gene.

From this rationale the likelihood of a mutation increases as the dose increases. If the father were exposed, more female embryos would be lost, and the sex ratio (males/females) at birth would rise in proportion to dose. If the mother was exposed, more male embryos would be lost, and the sex ratio at birth would fall in proportion to dose. If both parents were exposed, the resulting sex ratio or proportion of male births would be related to the individual parental doses and the frequency of dominant versus recessive lethal mutations. As can be seen, this theory of sex determination made fairly specific predictions that could be compared with the actual observations that were accumulating in survivors of the Hiroshima and Nagasaki atomic bombs.

When the data from the years 1948 through 1953 were examined, it appeared that the proportion of male births, in fact, declined with dose when the mother was exposed and increased, albeit modestly, with increasing paternal dose (Schull and Neel, 1958). The rate of change with dose was not, however, statistically significant, although it was in the direction predicted by the theory outlined above. For this reason, when the clinical phase of the studies ended, data on the sex ratio continued to be collected on the supposition that the rate of change might become statistically significant with further information. To this end, observations of the frequency of male births were continued through 1966. However, when these additional observations were analyzed, the results did not support the earlier findings; indeed, the modest changes seen were opposite those predicted by theory (Schull et al., 1966).

Today, the earlier understanding of sex determination is known to have been overly simplistic. First, it did not take into account the occurrence of X chromosome aneuploids, as in the Klinefelter and Turner syndromes, which make the predictions less precise. The first of these aneuploids was discovered in 1959 (in unrelated studies), and soon thereafter many others were identified. As a result of these discoveries, it is known that it is possible for some females to have only one X chromosome (or even as many as five X chromosomes), and for some males to have two or more X chromosomes. Moreover, these individuals with abnormal numbers of X chromosomes are more frequent in most populations than one would expect following exposure to ionizing radiation, at least at low doses. Second, it is now known that in females only one of the two X chromosomes within a cell is functionally active. This inactivation of one of the X chromosomes makes the prediction of the behavior of a potentially lethal gene on the X chromosome more difficult, particularly if the inactivation is not random (and it does not appear to be random). Given these developments, the simple, early arguments are much less compelling, and prediction of the effects of lethal mutations on the proportion of male births more tenuous. Thus, the sex ratio is no longer considered useful in estimating of genetic risks following exposure to ionizing radiation.

MORTALITY AMONG THE CHILDREN OF EXPOSED PARENTS

The largest study by far of the effects of parental exposure to ionizing radiation on mortality among their children conceived subsequent to irradiation is the study of the offspring of the atomic bomb survivors. That study involves the surveillance of approximately 72,000 children born alive between May 1946 and December 1984. It includes individuals whose parents were exposed at a wide variety of ages and doses. Surveillance of a cohort of this size is possible in Japan because of the existence of a unique record resource. Since the latter part of the 19th century, the Japanese government has maintained an obligatory system of household censuses, known as the *koseki*. These censuses, which are under the jurisdiction of the Japanese Ministry of Justice, incorporate information on all events that affect the composition of a family, such as birth, death, adoption, and marriage. It is thus possible to determine the life status of an individual wherever that individual may reside in Japan if the location of the individual's household record is known.

Customarily, on a cyclic basis, the *koseki* of the cohort are inspected anew to identify deaths that may have occurred since the last review cycle. If an individual has died, it is possible to obtain a copy of the death certificate and determine the stated primary cause of death and secondary contributors through the regional health centers of the Ministry of Health and Welfare. Follow-up is virtually complete; the vital status of more than 99% of individuals in the cohort can be determined. On those rare occasions when vital status cannot be determined, it is usually because of the migration of the individual to another country.

Insofar as malignant tumors are concerned, the mortality data are supplemented with information in the tumor registries maintained by the Medical Associations of Hiroshima and Nagasaki with the assistance of the Radiation Effects Research Foundation. These registries date from 1957 (Hiroshima) and 1958 (Nagasaki). The causes of death or incident cases are routinely coded by using the current International Classification of Disease and the International Classification of Cancer of 1976.

The first summarization of the mortality data occurred in 1966, and since then other publications have appeared, the most recent being in 1991 (Yoshimoto et al., 1991). At that time, the age of the average living member of the cohort was 28.8 years, and some 80% had completed their nineteenth year of life. The average dose received by the parents, father and mother combined, was 430 mSv (43 rem) on the basis of a neutron of 20 radiobiological effectiveness (RBE). It should be noted that the assumption of an RBE of 20 is not critical to what follows since the results described here are not materially affected by the choice of any other RBE between 1 and 20. Deaths were divided into those attributable to cancer and those attributable to other causes. Emphasis in the analysis of these data was placed on the 67,586 liveborn children when the parental doses could be computed by using the DS86 system (Shimizu et al., 1992) of dosimetry; however, a second analysis was based on all 72,228 children. That analysis used imputed doses received by the parents of the 4,642 children for whom direct calculation of the parental DS86 dose was not possible. The two analyses did not differ appreciably in their results.

Briefly, the findings were as follows. With regard to cancer, among the 26,894 liveborn children whose parents received DS86 doses of 10 mSv (1 rem) or more, there were 40 cases of cancer (1.5 per 1,000 births) and 14 instances of benign or unspecified tumors (0.5 per 1,000 births), and among the 40,692 children whose parents were not exposed or who were exposed to less than 10 mSv Sv (1 rem), there were 75 cases of cancer (1.8 per 1,000 births) and, again, 14 cases of benign or unspecified tumors (0.3 per 1,000 births) (Yoshimoto et al., 1991; see also Yoshimoto et al., 1990). Statistical analyses revealed no significant increase in the frequency of cancer deaths (cases) with increasing parental dose either for all cancers combined or for leukemia, the most common of the childhood cancers, which were considered separately. This was also true when the data were restricted to those individuals whose fathers alone were exposed. However, examination of the data suggested that only 3–5% of the tumors of childhood that were observed are associated with an inherited genetic predisposition that would be expected to exhibit an altered frequency if the parental mutation rate were increased. In other words, the mutational component was small.

With regard to causes of death other than cancer, 3,709 deaths occurred among the 67,586 individuals in the cohort with known parental DS86 doses (54.9 per 1,000 births). Again, the frequency of such deaths did not increase in a statistically significant manner with increasing parental dose. This was also true for four major disease categories, namely, infectious and parasitic diseases, diseases of the respiratory system, diseases of the digestive system, and certain conditions originating in the perinatal period. However, if the data on all diseases except neoplasms are taken at face value, there is a small but nonstatistically significant excess relative risk with increasing dose. If this excess relative risk of 0.038 per Sv (per 100 rem) for death by age 20 is accepted as real and not ascribable to chance, the increased risk of death per 0.01 Sv (per 1 rem) is 0.00038. To the extent that these values are applicable to the U.S. population, they imply that in 1983 when the probability of a liveborn child dying by age 20 in the general U.S. population was 0.0200, if the father had been exposed to 100 mSv (10 rem) this probability would be about 0.0238 (National Center for Health Statistics, 1986). Put somewhat differently, in 1983, when two individuals in every 100 live births would have been expected to die by their twentieth year of life, if the father had been exposed to 100 mSv (10 rem), the number of expected deaths would be about 2.4. This increase is too small to be within the resolving power of present epidemiologic studies.

CANCER AND LEUKEMIA IN PARTICULAR

U.S., Japanese, Russian, and Italian scientists have reported transgenerational cancer in the mouse or rat after exposure of the parent either in utero or before mating to carcinogenic chemicals or ionizing radiation (reviewed by Tomatis, 1994). Several years ago a report in the British media of an excess number of cases of childhood leukemia in the village of Seascale near the Sellafield nuclear facility in West Cumbria, prompted a more careful case–control study to ascertain whether this alleged excess could be explained. The findings of the resulting study by Martin Gardner and colleagues supported the earlier allegation (Gardner et al., 1990 a, b; Gardner, 1992). These investigators suggested that the excess was due to the relatively high doses received occupationally by the fathers of the patients. More specifically, they argued that paternal preconception exposure, in particular, exposure in the six months immediately prior to the conception of a child, induced mutations in sperm that resulted in the offspring developing leukemia. Their findings stimulated a number of other studies aimed at confirming, if possible, their findings.

Little or no support for the hypothesis that paternal exposure leads to an excess risk of childhood leukemia has been found in studies carried out in the United States (Jablon et al., 1991), France (Hill and Laplanche, 1990), Germany (Michaelis et al., 1992), and Canada (McLaughlin et al., 1993). Moreover, the findings among children at Sellafield were significantly at variance with those among the children of the atomic bomb survivors (Little, 1990, 1991, 1992). A further complication in the acceptance of the findings and hypothesis of Gardner and colleagues has been the finding of similarly raised levels of leukemia around potential nuclear facilities in Britain (Cook-Mozaffari et al., 1989) and Germany

(Michaelis et al., 1992). This illustrates anew the difficulties inherent in attempting to interpret any cluster of cases. Nonetheless, the issue remains contentious, and at least one other multi-institutional case–control study has reported findings that suggest an increased risk of infant leukemia among the children of fathers exposed to diagnostic X rays prior to the conception of the child (Shu et al., 1994). That study, however, could not confirm the effect of maternal exposure prior to conception that had been reported by Bithell and Stewart (1975), nor did its design permit the assessment of the dose–response relationship for exposed fathers.

IMMUNE DEFICIENCY

Various genetic abnormalities cause severe immunologic deficiencies in the offspring, and these deficiencies predispose them to infection and non-Hodgkin's lymphoma. Among these diseases are congenital agammaglobulinemia, ataxia-telangiectasia, Wiskott-Aldrich syndrome, Bloom syndrome, and Chediak-Higashi syndrome (Seibel et al., 1993). An excess of such syndromes, infections, or lymphomas has not been observed as causes of mortality in the F_1 generation among survivors of the · atomic bomb blasts in Hiroshima and Nagasaki (Yoshimoto et al., 1991).

NEUROLOGIC DEFICIT, INCLUDING MENTAL RETARDATION

Definitions of the terms "neurologic deficit" and "mental retardation" vary widely. This is perhaps inevitable when there is a continuum of intellectual or neurologic potentials. Still, it complicates the comparison and integration of the findings from different studies when the definition has not been the same among studies. For example, among the survivors exposed in utero to the atomic bombing of Hiroshima and Nagasaki, an individual was considered to be severely mentally retarded if he or she was unable to form coherent sentences, to perform simple arithmetic tasks, or to manage his or her own affairs or was institutionalized. The World Health Organization restricts the term "severe mental retardation" to those individuals with an IQ of less than 50; individuals with IQs in the range of 50 to 70 are described as mildly retarded.

Diagnosis of mental subnormality may rest on the clinical experience of the examining physician, on structured tests of mental performance, or commonly, on both. The frequency of the diagnosis depends on the severity of the handicap and the age, at least through the school years, at which the individual is examined (Gesell and Amatruda, 1975). Subtle limitations of mental performance that are readily recognizable in the pubertal child may be very difficult to diagnose in a newborn. This difference is well illustrated by the findings of the Collaborative Perinatal Project conducted by the National Institute of Neurological

Diseases and Stroke. That study, which began on January 1, 1959, and ceased registration in December 1965, involved 15 university-affiliated medical centers and led to the study of 55,908 pregnancies (DHEW, 1973). Although the study had a number of objectives, one of the major aims, as initially stated, was to "determine the relationship between factors in the perinatal environment and the continuum of human reproductive failure, with particular reference to the central nervous system for: (a) early manifestation of deficits (infancy and early childhood), and (b) later manifestations of deficits (5 to 15 years)."

Although this study remains one of the better sources of information on the frequency of congenital malformations, including neurologic deficits, available for the population of the United States, it also illustrates that, since a neurologic handicap is not a single disease entity and the potential causes of a deficit are many, it can be extremely difficult to recognize the cause in a specific instance. It is known that inherited gene and chromosomal defects, as in Down syndrome or the fragile X syndrome, can lead to the occurrence of mental retardation, but attempts to estimate the contribution that genetic factors make to the overall frequency of mental retardation have generally been unsatisfactory. Most instances of mental retardation are idiopathic; that is, they have no clear identifiable cause.

The situation is different, however, with respect to the exposure of a pregnant woman to noxious agents, particularly early in her pregnancy, when untoward effects on the mental development of the child have been demonstrated. Among those agents for which the evidence is most persuasive are ionizing radiation, methyl-mercury, lead, and alcohol. The effects of exposure in these instances can manifest themselves in a variety of ways: a possible increase in the frequency of overt mental retardation, a loss in performance on standard intelligence tests, and poorer performance in school. The findings, particularly with respect to ionizing radiation, have been reviewed in some detail in the 1993 report of the United Nations' Scientific Committee on the Effects of Atomic Radiation (UNSCEAR, 1993). These effects are not genetic in origin, although the possibility that individuals may differ in their sensitivities to noxious agents for genetic reasons cannot be rejected. The effects stem from the exposure of the developing embryo or fetus itself to the noxious agent at vulnerable stages in its development.

Unfortunately, little evidence dealing with the risk of mental retardation among infants conceived following parental exposure to ionizing radiation is available. Clinical studies of the children of the atomic bomb survivors conceived after the exposure did not extend beyond the 10th month following birth and thus provide limited information. The data that are available from the study of atomic bomb survivors fail to demonstrate an increased risk of any form of mental retardation, including an increased risk of Down syndrome.

8

Feasibility of the Study of Adverse Reproductive Outcomes in the Families of Veterans Exposed to Ionizing Radiation

The feasibility of a study of adverse reproductive outcomes among the families of veterans exposed to ionizing radiation hinges largely on the answers to five interrelated questions. First, how is a suitable sample or cohort (numerators and denominators) to be defined, and can this be done without inadvertently introducing selection biases that could obscure a true effect or produce a spurious one? Second, what will be the probable size of that sample or cohort? And, as a corollary, will that size be large enough to reveal effects of the magnitude anticipated on the basis of current knowledge? Third, what is the probable dose distribution among the members of that sample or cohort? Fourth, how reliable are the individual dose estimates? Fifth, what mechanisms are available for identifying adverse reproductive outcomes? Each question will be considered separately.

DEFINITION OF A SUITABLE SAMPLE OR COHORT

Anecdotal information can be valuable in establishing the need for an epidemiologic study, but self-volunteered information is unlikely to provide a basis for reliable estimates of risk since experience shows that persons with a personal or even financial interest in an exposure to some hazard will selectively respond. Accordingly, a scientifically defensible and valid study of the effects of ionizing radiation on reproductive outcomes depends on the availability of a representative sample of exposed veterans and their families, and the means to establish these outcomes without reference to whether they were normal or abnormal. The Nuclear Test Personnel Review (NTPR) program of the Defense Nuclear Agency (DNA) has identified some 210,000 veterans who participated in one or more atmospheric tests involving the detonation of a nuclear weapon. Those individuals or a suitably large and representative sample might provide the basis for a study cohort, and it seems probable that deaths among these veterans could be determined through the records of the Department of Veterans Affairs or other sources. However, it is far more difficult to trace an unbiased sample of living persons, given the lack of identifying information in the original records. Furthermore, the available records do not contain information on the reproductive histories of the veterans, that is, their children, estimated to be about 500,000 in number, and grandchildren, and for reasons adduced elsewhere in this report it seems doubtful that such information can be reliably and accurately obtained at this late date. Thus, the committee concludes that, whereas a study of the life status and health problems of the veterans themselves is feasible (and is in fact being done), the means do not exist to obtain information on adverse reproductive outcomes among their children and grandchildren in a suitably complete and unbiased manner to estimate the risk, if any, stemming from exposure to ionizing radiation.

SIZE OF THE SAMPLE OR COHORT REQUIRED

To determine the size of an epidemiologic study seeking to respond to the concerns of the Atomic Veterans, two somewhat different, but related questions can be posed:

1. If current estimates of risk are correct, how large a sample would be needed to demonstrate that risk?
2. Given the sample size that might be available, how large would the risk have to be to be demonstrable?

The committee has sought answers to both of these questions. The results follow.

Question 1

Table 7 provides a comparison of certain characteristics, relevant to the feasibility of an epidemiologic study, of the Japanese atomic bomb survivors and their children with the Atomic Veterans and their children. The main differences are average dose and dose range, the number of children potentially available for study, and the environmental conditions existing after birth. Although the potentially larger sample of children from the Atomic Veterans is favorable to an epidemiologic study, the much lower average dose and limited range of doses more than offsets this advantage in numbers.

TABLE 7. Comparison of Characteristics of Hiroshima–Nagasaki Atomic Bomb Survivors with Those of Atomic Veterans

Characteristic	Hiroshima–Nagasaki Survivors	Atomic Veterans
Average dose (Sv)	0.44 (conjoint)	0.006
Dose range (Sv)	0–4.0 (conjoint)	0–0.03[a]
Number of children	52,000	500,000
Parent exposed	11,000 both parents	Father only
	18,000 mother only	
	6,000 father only	
	17,000 neither parent	
Births	At home	In hospital
Postnatal environment	Adverse	Normal conditions

[a] A total of 95% of the Atomic Veterans received doses in this range.

In discussing the feasibility of studying the children of the Atomic Veterans in order to estimate the genetic effects of radiation exposure, it may be useful to rewrite Equation 1 (page 31) as follows:

$$(S + I)/S = 1/DD \times MC \times D + 1 \; . \qquad \text{(Equation 3)}$$

This form gives the expected total number of cases in the exposed group $(S + I)$ divided by the number expected without the added exposure (S). This is the relative risk, RR (see Chapter 2).

The genetic risks associated with low doses of ionizing radiation are quite small when compared with the background incidence of genetic disease in the general population. To give the reader a feel for the magnitude of these risks, three quantities are estimated: (1) an upper-bound relative risk for major congenital abnormalities in the children of all Atomic Veterans, (2) a best-estimate relative risk for genetic disease in general in the children of all Atomic Veterans, and (3) a best-estimate relative risk for genetic disease in general in the children of Atomic Veterans who were exposed to 100 mSv (10 rem) or more.

What is the maximum relative risk one could expect to find among the Atomic Veterans' children given what is known about the effects of ionizing radiation from studies in animals and humans? Consider, for example, the appearance of major congenital malformations among the children of exposed Atomic Veterans. From the data on mice and humans reviewed above, the doubling dose is certainly expected to be no less than 250 mSv (25 rem) (100 rem is the usual estimate for a low dose). Assume a mutational component of 1/2 (BEIR V [NRC, 1990] assumed a maximum mutation component of 0.35 for this endpoint) and assume that 1/10 of the effect will appear in the first generation following exposure (again, this is greater than the most probable value). In addition, assume that, on average, the Atomic Veterans were exposed to 20 mSv

(2 rem) (the current estimate is 6 mSv [0.6 rem]). Finally, the fact that only fathers were exposed must be taken into account. The estimate of the maximum relative risk (RR) to be expected is then

$$RR = (S + I)/S = (1/25)(1/2)(1/10)(1/2)(2) + 1 = 1.002 .$$

This is a very small relative risk. It means that the increase in the incidence of major congenital malformations due to the ionizing radiation exposure is expected to be only 0.002 of the background incidence normally seen in the population. This is an upper limit based on extreme assumptions that were all chosen to overestimate the risk. Suppose the 210,000 Atomic Veterans had a total of 500,000 children. Without any radiation exposure to the fathers, about 15,000 of these children would be expected to be affected by a major birth defect; with an additional average exposure of 2 rem the expectation, under these extreme assumptions, is 15,025 affected children. In fact, the BEIR V "best" risk estimate for major congenital malformations implies a relative risk of only 1.0004 in the first generation, that is, only five additional cases of malformations due to the radiation.

BEIR V (NRC, 1990) estimated genetic risks from ionizing radiation for seven kinds of disorders: clinically severe and mild autosomal dominant, X linked, autosomal recessive, unbalanced translocations, trisomies, and congenital abnormalities. For all seven endpoints combined the relative risk from the BEIR V risk estimates in Table 2-1 of that report is less than 1.0006 for first- generation effects (assuming 1 rem of exposure to the male). Again, suppose that the 210,000 Atomic Veterans had a total of 500,000 children. Without any radiation exposure to the fathers, about 21,150 of these children would be expected to be affected by some disorder in these seven categories; with an additional average exposure of 1 rem the expectation is 21,164 affected children—an increase of only 14 children.

Finally, consider the relative risk for the children of Atomic Veterans who received 10 rem or more. According to present dose information, this 0.07% of the total cohort has an average dose of approximately 20 rem. For the genetic risks in general (seven endpoints combined) the expected relative risk for this subset of Atomic Veterans—the most heavily exposed—is 1.012. Suppose that these men had 350 children. Without any radiation exposure to the fathers, about 15 of these children would be expected to be affected by some disorder in these seven categories; with the additional 20 rem the expectation is 15.18 affected children—less than one additional child.

A similar calculation can be made using the "direct method" explained in Chapter 6. In Table 1 of Appendix G of UNSCEAR 1993, the incidence of genetic or partially genetic disease having serious health consequences before the age of 25 years is estimated to be approximately 79,400 per million live births. UNSCEAR estimates an additional 15–30 seriously affected individuals per million born to fathers exposed to 0.01 Gy of low-LET radiation. Taking thirty as an upper estimate, multiplying by 3 to apply to a high dose rate response as a worst case, and multiplying by 20, about 1,800 additional seriously affected offspring per million children in the first generation following exposure of their fathers are expected. This corresponds to a relative risk of approximately 1.023. Thus of the estimated 350 children born to fathers exposed to an average of 20 rem, 27.79 would be expected to be affected without the radiation and 28.43 with the additional radiation. Again, less than one additional case would be expected due to the radiation.

The committee did not count the large genetic disease burden usually referred to as diseases of complex etiology, but these are expected to have relative risks at least as small as the ones considered here. Also, the risks to the grandchildren of the Atomic Veterans will be even smaller.

It should be clear from the examples given here that effects of this magnitude are not measurable in any attainable sample, because the induced cases make up a miniscule part of the spontaneous burden of human genetic disease and are indistinguishable from the naturally occurring cases.

If this estimate of the probable relative risk is correct, or nearly so, the sample size needed would run into several millions of children (or grandchildren), a number that is certainly beyond the limits of feasibility for an epidemiologic study. This can be shown more formally. To do so, however, the committee digresses briefly for some background remarks.

When a statistical test is performed, it is done in the context of a hypothesis. Commonly, that hypothesis is known to statisticians as the null hypothesis; it postulates that there is no difference between the groups under study. On the basis of the results of the test performed, that hypothesis is either not rejected or rejected. If the null hypothesis is not rejected, it is equivalent to asserting that any difference between the two groups may be due to chance. If the null hypothesis is rejected, the difference between the groups is possibly a consequence of the exposure under study. For example, if a group exposed to ionizing radiation was compared with one not exposed and the null hypothesis is rejected, it is possible to conclude that radiation may be associated with the health outcome. The likelihood of observing an association between the exposure and the outcome depends upon the sample sizes involved, the magnitude of the difference between the two groups, and the errors of interpretation of the data an investigator is willing to accept. These errors are of two types. First, the investigator may reject the null hypothesis, that is, conclude that the difference is probably not due to chance when, in fact, it is. This is commonly referred to as a type I error, and in computing the sample size needed to demonstrate a particular risk one must set the acceptable probability of such an error (usually 5 or 10%, or more, generally designated α). Second, the investigator may fail to reject the null hypothesis (no difference between the compared groups) when, in fact, there is a difference between the groups. This is termed a type II error, and the probability of such an error is often set at 20% (and is designated β). It warrants noting that the complement of this latter error rate (i.e., $1 - \beta$) is called the power of the test, which is 80% in the example.

Table 8 sets out the sample sizes required to demonstrate specific relative risks assuming different background frequencies of an event and error rates of 10% (type I) and 20% (type II).

There is a functional relationship between the sample size, the difference that is obtained between two (or more) groups under comparison, and the frequencies of the two types of errors. This relationship is such that if any three of these four values are known (or the investigator is prepared to assume their values), the fourth is also known. By using this relationship, it is possible to compute that fourth value. For example, if one sets the frequencies of the two types of errors and the relative risk (which is the difference between the two groups of interest), one can compute the required sample size. Table 8 sets out the results of such computations in the present context. The specific methodology (Statistics and Epidemiology Research Corporation, 1993) used is based on the detection of a statistically significant trend (dose–response) test assuming relative risks from 1.5–4.0 for the highest dose category (>100 mSv [>10 rem]) and for various outcome frequencies.

TABLE 8. Sample Sizes Required to Detect a Range of Relative Risks for the Highest Dose Category >100 mSv (>10 rem) in a Test for Trend[a]

Frequency of Outcome(%)	Sample Size for Relative Risk of:			
	1.5	2.0	3.0	4.0
0.1	2,816,000	868,000	329,000	180,000
0.5	580,000	182,000	64,000	36,000
2.0	148,000	47,000	16,000	9,000
3.0[b]	100,000	32,000	11,000	6,000

[a] The estimated sample sizes were calculated for those in the highest dose category (>10 rem), with type 1 and type 2 errors of 10 and 20, respectively. The sample sizes have been rounded to the nearest thousand to avoid an undue perception of accuracy.

[b] Frequency of major congenital defects in the general (unexposed) population.

From Table 8, assuming a background rate of 3% for major congenital birth defects present at birth, a total study population of 100,000 individuals (unexposed and exposed) would be required to detect a relative risk of 1.5. Therefore, if the relative risk is, in fact, considerably less (i.e., 1.002), the sample size would be in the millions.

Question 2

If the circumstances are such that an investigator has little control over the size of the study groups, then he or she might ask what difference could be demonstrated with a particular sample size. The computations are very similar to those underlying Table 8, but now the sample size and the frequencies of the two types of errors alluded to above are assumed to be known and it is the relative risk that is to be calculated. In Table 9 the committee presents a series of such calculations for various adverse reproductive outcome frequencies (f) and sample sizes (N).

TABLE 9. Minimum Detectable Relative Risk for a Range of Sample Sizes for the Highest Dose Category >100 mSv (>10 rem)[a]

Frequency of Outcome (%)	Relative Risk for Sample Size of:			
	80,000	140,000	210,000	500,000
0.1	6.7	4.7	3.6	2.5
0.5	3.7	2.2	1.9	1.5
1.0	2.1	1.7	1.6	1.4
2.0	1.7	1.5	1.4	1.3
3.0[b]	1.5	1.4	1.3	1.3

[a] The estimated relative risks were calculated for those in the highest dose category (>10 rem), with type 1 and type 2 errors of 10 and 20, respectively.

[b] Frequency of major congenital defects in the general (unexposed) population.

Again, assuming a population of 500,000 and a background rate of 3% for major congenital defects present at birth, the smallest detectable relative risk would be 1.3, which is 150 times greater than that presumed to be likely (1.002). The total sample size that would be required to demonstrate a maximum relative risk of 1.002, assuming a 3% frequency of outcome, is approximately 212,000,000.

DOSIMETRY OF ATOMIC VETERANS

Radiation dose is generally considered in two parts, external and internal. External dose is that received from a radiation source outside the body such as radioactive materials on the ground or on equipment such as vehicles. For the exposed Atomic Veterans, the biggest components were gamma rays and beta rays, but only a small fraction of the beta radiation could penetrate the clothing and skin to reach the gonads, the specific organ of interest here.

The candidate database for radiation doses for the Atomic Veterans is the Nuclear Test Personnel Review (NTPR) program established in 1978 by the Defense Nuclear Agency (DNA). The work of assembling the dose information was accomplished by the military branches and by contractors working for DNA. This work continues with the objective of obtaining the best estimates of dose for as many veterans as possible. Because only limited measurements were made for some test operations, particularly before 1955, the NTPR database includes doses that were estimated from few data and with extensive use of models. The models are mathematical relationships based on extensive measurements made during later weapons tests and laboratory studies. Although a few veterans were exposed to neutrons, the bulk of the recorded doses were from gamma radiation. In some cases there was the potential for materials containing radioactive particles to be inhaled or ingested, and efforts were made to include doses from these particles in the dose estimates.

In some cases, the major component of the dose received, gamma radiation, was measured with film badges. In other instances, dose was inferred from the badge worn by a companion performing the same activities and at the same location. In others, it was reconstructed from measurements made in the radiation field with instruments and from the time spent by the veteran in that radiation field. In many instances, the dose was estimated only for a relatively large group such as a platoon, a crew of a boat, or a work party. The less specificity available for the estimation of dose, the greater the uncertainties in that estimate. Uncertainty in some of the dose estimates, especially for those people who were not badged, is unavoidable. Unfortunately, for a large number of veterans, the doses must be estimated from very little information, and the accuracies of the doses are correspondingly poor.

For many people, a cause for concern has been the magnitude of the dose due to internal emitters such as plutonium. Once radioactive materials are inside the body, types of radiation that are essentially harmless when they are on the outside, such as alpha and beta particles, can irradiate cells, tissues, and organs. An accurate assessment of doses to the gonads is more difficult for some of these materials.

Internal doses were found in early tests to be a small part of the total dose in general, but the potential for radioactive materials to be taken into the body by inhalation or ingestion may have existed in some cases, adding to the uncertainties about the total dose. All of these factors and others have been studied for many years. The *Effects of Nuclear Weapons* (Glasstone, 1962) presents discussions of internal and external dose. For example, Chapters VIII and XI of Glasstone (1962) discuss this subject as it was understood in 1962, and NRC (1985) discusses this subject from the vantage point of the Atomic Veterans.

An NRC report (1985) concluded that uncertainties about the internal dose are large for the Atomic Veterans but that the overall doses were small compared with the external levels of gamma radiation. The highest potential for internal exposure appears to have existed for the 28 men stationed on Rongerik atoll at the time of the Castle Bravo test. The 1985 NRC report gives estimates of the doses that these men received and concludes that the highest doses were to their thyroids. The doses to their gonads were not estimated, but the gonads would have been affected little by internally deposited radionuclides.

In another evaluation (IOM, 1995), the Institute of Medicine's Committee to Study the Mortality of Military Personnel Present at Atmospheric Tests of Nuclear Weapons reviewed the NTPR database and the methods used by NTPR to estimate doses. The review found the NTPR dose data to be unsuitable for dose–response analysis. However, the committee believed that comprehensive dose reconstructions may be feasible for a limited subset of Atomic Veterans.

Although the DNA dose data are unsuitable for dose–response analysis, they may provide a rough estimate of the magnitude of doses received by the Atomic veterans. Of 210,000 participating veterans, about 1,200 received doses that were estimated to exceed 50 mSv (5 rem) (DNA, 1995a), which is the present annual exposure limit set by the U.S. Nuclear Regulatory Commission (10CFR20) for workers occupationally exposed to radiation. About 20,000 participants (DNA 1995b) have assigned doses that exceed the more conservative annual occupational limit, 20 mSv (2 rem), proposed by the International Commission on Radiological Protection (1991). A total of 0.07% of the doses exceeded 100 mSv (10 rem), and the average dose for the Atomic Veterans was 6 mSv (0.6 rem). Although the dose assigned to a given veteran might change with further study, the distribution of doses across the cohort is unlikely to change significantly.

Two groups of veterans require additional comment: the veterans who entered Hiroshima and Nagasaki at the beginning of the occupation of Japan to assist in the cleanup and prisoners of war who may have been taken into the two cities on work parties.

Because of the interest of the scientific community in the potential exposures that were received by early entrants in Hiroshima and Nagasaki after the atomic bombings, studies were done in the early 1960s to determine whether or not these exposures were significant from a biologic point of view. The term "early entrants" denotes those persons who were not close enough to the detonation to receive a dose directly from the bombs, but who walked into the ground zero area within hours to a few days after the detonations and thus got some exposure to either fallout or activation products. Because Japanese scientists made radiation measurements within a few days of the bombings and both American and Japanese scientists made measurements about a month later, there is a body of information on which to make comparisons with theoretical measurements and other measurements made at later weapons tests.

Arakawa (1962) reported an extensive study of potential doses to individuals who may have entered the ground zero and fallout areas. His data show that people entering either the ground zero area or areas of maximum fallout at the time that U.S. forces landed in Japan would have been in the millirem-per-hour range as a maximum. In fact, his studies show that residents of Nishiyama (an area about 3 kilometers to the east of the hypocenter where the bulk of the fallout in Nagasaki occurred) would have received the highest doses of any people not directly exposed to the bombs and that their lifetime doses, assuming that they never left the area, were below the level at which biologic effects would be detectable.

More recent dose reconstructions on occupation forces in Hiroshima and Nagasaki estimated upper-bound doses based on worst-case scenarios (DNA, 1980). The external doses ranged from 0.3 mSv (.03 rem) at ground zero in Hiroshima to 6.3 mSv (.63 rem) in the Nishiyama area near Nagasaki. Whole-body internal exposures ranged from 0.03 mSv (.003 rem) in Hiroshima to 0.68 mSv (.068 rem) in the Nishiyama area. Therefore, it is not likely that people entering any area of either Hiroshima or Nagasaki in September 1945 would have received a dose of as much as a 10 mSv (1 rem) and that a casual visit to the area would have caused an exposure in the range of only a few millirem.

The final group of veterans to be considered is the one composed of U.S. prisoners of war. There is no record of any prisoners being held in the Hiroshima area, and no one claims to have been held there, so the main concern is for persons who may have been held captive near Nagasaki and who may have been taken into the city on work details. It is known that a few prisoners were held north of Nagasaki, and some of these say that they were used on work parties in the city after the bombing. In this case, Arakawa's early-entrant calculations for Nagasaki would be most appropriate. The biggest contributor to exposure rate was sodium-24, which has a half-life of less than 15 hours, meaning that after 15 hours half of it has gone away by decay, after 30 hours only one-fourth remains, and so forth. If a work party entered the ground zero area of the city the day after the bombing, which under the conditions of communications, management structure, and so forth, seems unlikely, and if the work parties returned daily for a full work day until their release, their doses would have been very low, probably less than 10 mSv (1 rem). Such doses and the limited number of people involved would make this an unlikely basis for an epidemiologic study.

There has been no statistically significant demonstration in the populations of Hiroshima and Nagasaki of any induced hereditary effects of radiation. With regard to veterans, within the constraints of the uncertainties, it is clear that the average dose as well as the highest measured dose to veterans were small compared with the minimum doses at which the effects of concern were possibly observed in other exposed populations such as the survivors of the atomic bombings of Hiroshima and Nagasaki. The organ of concern for this study is the gonads, but within the other uncertainties in doses, the dose to the gonads can be taken to be the same as the estimated dose to the whole person.

IDENTIFICATION OF ADVERSE REPRODUCTIVE OUTCOMES

Study of reproductive outcomes among Atomic Veterans requires being able to identify both normal and abnormal outcomes in an unbiased manner. Although a nonconcurrent cohort approach, which identifies groups of veterans who differ with regard to radiation exposure but are otherwise similar and follows them forward in time to determine if rates of reproductive outcomes differ by exposure group, would seem to be a logical approach, it is probably not feasible. These groups are likely to have completed their families at least 15 years ago, and the records necessary to identify adverse reproductive outcomes during a time period of from 15 to 50 years ago are not likely to be available. During this time there have been major changes in the information recorded in the vital records. For example, information on variables such as birth weight and gestational age has not always been required to be recorded on the birth certificate in every state. In addition, some states do not maintain the medical information reported on the confidential portion of the birth certificate, for example, information on congenital malformations, with personal identifiers.

Reporting of fetal deaths varies from state to state. For example, many states require that only those fetal deaths occurring at 20 weeks of gestational age or greater be reported, whereas other states require that all fetal deaths, regardless of age be reported. There is known to be marked underreporting of fetal deaths at early gestational age, the stages at which the rates of loss are the highest.

In addition to underreporting of fetal deaths, underreporting of early neonatal deaths in very low birth weight newborns has been observed. This has been recognized as a problem in some areas during the last decade and is likely to have been even more prevalent earlier. Even for infant deaths by other causes, there is great variability in the accuracy and completeness of recording of causes of deaths.

To study the health conditions of greatest concern and health outcomes for which a biological mechanism related to paternal exposure can be postulated, unbiased information is even less likely to be available than for other types of adverse reproductive outcomes. This would include congenital malformations that are known or believed to include a genetic component in their etiology. Although it may be possible to obtain some information from Atomic Veterans or their spouses on their children who may have had birth defects, the medical and vital records necessary to validate this information in an unbiased way are not likely to be available. If a complete cohort of births to the wives of Atomic Veterans could be identified, then, in theory, it might be possible to determine birth defects diagnosed in the newborn period by reviewing the babies' medical records, but a large percentage of such records are likely to be unavailable because of the length of time that has passed since the events occurred. Hospitals will have closed or changed ownership, records will have been purged or destroyed, and records will simply have been lost. Because of the size of the cohorts of interest and their geographic dispersal, it would be highly infeasible to determine the existence of medical records, let alone their availability or completeness. The challenges of studying other outcomes, such as spontaneous abortions, learning disabilities, and mental retardation, would be even greater. These latter endpoints are difficult to study epidemiologically in defined contemporary populations and would be nearly impossible to study adequately in a historical cohort.

Potential for recall bias is a particular concern for some endpoints, such as spontaneous abortions. Spontaneous abortions present a number of methodological problems for study in contemporary populations (Sever, 1989). It would be extremely difficult to study them in an unbiased way in populations that were at the height of their reproductive lives more than 30 years ago. In groups of women who have been questioned about their history of spontaneous abortion, recall seems to be relatively accurate for the period up to 20 years prior to the interview, however before that time recall is poor on the basis of a comparison of contemporary reports with later recall (Wilcox and Horney, 1984). This is in the absence of any concern about a potential association with an exposure that might lead to reporting or recall bias (White et al., 1989).

For many of the health outcomes of interest there is the potential for marked variability in the diagnostic criteria used in different years and in different areas. As noted earlier, mental retardation represents a wide variety of possible diagnoses that share the common feature of some decrease in mental ability. Mental retardation is usually defined as an IQ of less than 70. Some definitions also include a functional component (Grossman, 1977). The prevalence of mental retardation has been shown to be related to age. For example, the relation between age and prevalence of mental retardation was shown in a cohort of 10-year-olds, in accordance with the belief that by age 10, such children would have been identified and could be ascertained through educational facilities (Yeargin-Allsopp et al., 1990).

Study of mental retardation in the children of Atomic Veterans would require access to school records that include information on standardized test scores. Such records are unlikely to be available. In addition, there are a number of other risk factors such as alcohol use by the mother during pregnancy or the mother's educational level that would be very difficult to control for in a historical cohort study.

9

Alternative Approaches

Data on the occurrence of adverse reproductive outcomes following exposure to ionizing radiation could be derived from a variety of cohorts, in addition to the atomic bomb survivors, such as the children of (1) people residing in areas where the background of naturally occurring radiation is substantially higher than usual, (2) individuals, other than Atomic Veterans, exposed to fallout from atmospheric weapons testing, (3) people living near nuclear installations, (4) individuals exposed occupationally, (5) patients undergoing medical diagnostic procedures, and (6) patients undergoing medical therapy for benign or malignant disease. Each of these cohorts has strengths and limitations. Usually, these are related to sample size, population composition, certainty of dose, presence of concurrent disease, and other confounding factors.

The study of the atomic bomb survivors is the largest, longest, and most comprehensive epidemiologic study of radiation-induced carcinogenesis and mutagenesis that has been undertaken. Its strengths are that it includes a large population of all ages and both sexes who were not selected because of occupation or disease. Other strengths are that it includes a wide range of doses, has included follow-up for more than 45 years, has comprehensive individual dosimetry, and can use internal comparisons. Weaknesses include the following: although the clinical examinations of the children of the survivors began in the spring of 1948 and the surveillance of mortality among these children covers the time since May 1946, the cohort on which the studies of cancer among the survivors themselves rests was defined on the basis of the 1950 national census and thus does not include the years 1945 through 1949. Moreover, of importance in the present context is the fact that the exposure was at a high rather than a low dose rate, and the possible contribution of neutrons is somewhat uncertain. The fact that the population is Japanese raises some question about the transfer of risk factors derived from this population to other populations that may have different baseline rates of health outcomes.

There have been a number of studies of populations (other than Atomic Veterans) exposed to radioactive fallout from weapons testing, weapons use, and nuclear plant accidents. These studies may involve relatively few people (as in the Marshall Islands), but most involve thousands or hundreds of thousands of people. The advantages of such studies is that they involve populations of both sexes and all ages. In addition, they may yield information on the effects of chronic exposure. The difficulties in risk assessment arise from the fact that the doses are usually quite low and are rarely available on an individual, specific basis. Typically, dose estimates are derived from computer modeling of the source, meteorology, environmental pathways, and assumptions about ingestion patterns and amounts. Often, the dose estimates can be made only collectively. Although fallout patterns can be modeled by computer, experience from the Chernobyl nuclear power plant accident in 1986 has shown that individual doses may vary by a factor of 10 or more from the estimated average.

Studies of fallout within the United States as a result of weapons testing at the Nevada Test Site have been performed as part of a significant scientific effort to reconstruct doses. The advantages of that study were comprehensive exposure evaluation and protracted exposures at a low rate. In spite of the dosimetry estimates, there still remains considerable uncertainty about individual doses. In addition, the estimated cumulative doses are much lower than those experienced from natural background radiation.

Studies have been performed on localized fallout from the 1954 BRAVO weapons tests in the Marshall Islands. Fewer than 8,000 people on these islands were affected. The advantages of that study are that the population was unselected, there has been an attempt at determining individual dosimetry, and there has been long-term comprehensive medical follow-up. The small sample size remains a problem for conducting studies in this population, as does the uncertainty about the dose due to short-lived radioiodines.

There have also been studies on the populations exposed to fallout as a result of the Chernobyl accident in 1986. The advantages of this group are that it is large, the population is unselected, and dosimetry has been done for highly contaminated villages. The limitations are that the length of follow-up is limited and the iodine dosimetry remains somewhat uncertain.

Data on the environmental contamination of the Techa River in the eastern part of Russia and Semipalatinsk, the weapons testing site of the former Soviet Union, have also recently become available. Difficulties in both instances include accurate estimation of the absorbed doses or inadequate ascertainment of exposed individuals. Strengths include a wide range of doses, long follow-up, an unselected population, and a large population.

Studies of populations living around nuclear power plants have the advantage that they can be well defined assuming that there has been little population mobility. Unfortunately, in the United States mobility is common and so an effort must be made to guarantee that the patients with disease, in fact, lived near the plant at the appropriate time before the latent period for cancer induction. The very low doses from emissions of most normally functioning nuclear power plants makes the required sample sizes for statistical significance almost impossible to achieve in circumstances where the exposure (dose) is less than that at Chernobyl.

Occupational studies are a major source of epidemiologic information. They have the advantage that work records and times of employment are known. Knowledge of the dosimetry in these studies ranges from good for workers in nuclear facilities to very poor for groups such as the early uranium miners. Another advantage is that there is usually a large number of people who can be studied. One problem with occupational studies is that the workforce is predominantly young or middle-aged healthy males, and the applicability of these risk factors to other populations requires some assumptions. The so-called healthy worker effect needs to be considered. This effect tends to give standardized mortality rates (SMR) or standardized incidence rates (SIR) that are less than unity. Confounding factors, such as exposures to chemicals and other substances in the workplace, also need to be considered. Smoking is another common confounding factor that needs to be considered. Moreover, in these studies, and indeed in all studies in which the dose is low, other sources of exposure to ionizing radiation, such as diagnostic or therapeutic irradiation, loom large as possible sources of confounding.

Exposures to ionizing radiation as a result of medical diagnostic procedures are another potential source for information about the effects of ionizing radiation. The advantages of these studies are that the doses are reasonably well known, as is the field irradiated. The doses may not be as precise on a percentage basis as those from radiation therapy since the doses are known to be low and the technical factors are not usually recorded. The exposed populations consist of people of both sexes as well as most age ranges. The generally low doses used in medical diagnostic procedures require extremely large sample sizes for statistical significance to be achieved. The purpose of the diagnostic study may also be a confounding factor, although generally the patients are available for long-term follow-up.

Studies of patients who have received radiation therapy for benign diseases have the advantage of a known exposure field, a known type of radiation, and usually, good dosimetry. However, the dosimetry can be a problem in some cases in which children were irradiated and the organ of interest was near the primary radiation field. In those cases, patient motion could cause substantial uncertainty in the actual doses to the organ of interest. The high doses of radiotherapy allow risk estimates to be derived with relatively small population groups, but the high doses also introduce the confounding factor of potential cell killing. The disadvantages typically are patient selection bias or confounding by disease, confounding by other therapies or economic status, and potential loss to follow-up if the disease did not require further treatment.

Studies of the treatment of malignant diseases with ionizing radiation have the advantages of a well-defined dose and a known type of radiation in a localized field, confounders can be documented, and good follow-up is possible because patients who receive these treatments are often followed closely for life. The disadvantages are the possibility of concurrent or other therapies that may confound the analysis, selection bias due to the disease, the possibility of cell killing rather than cancer or mutation induction, and a shortened lifespan for follow-up as a result of the malignant disease. A group including individuals who have been treated for malignant disease with ionizing radiation are survivors of childhood cancer. Current data do not indicate an increased risk for adverse reproductive and developmental effects in the offspring of male cancer survivors (Hawkins, 1991). A study of more than 20,000 survivors and their offspring currently being conducted in the United States and Canada should provide important new data on this particular population.

Other exposed populations, however, have not been studied or have been studied inadequately. These include the children of (1) medical personnel who are occupationally exposed, (2) workers in nuclear facilities, and (3) members of the armed forces whose service functions involve exposure to ionizing radiation such as nuclear submariners and the crews of aircraft of the U.S. Strategic Air Command (SAC).

Some of these groups offer opportunities pertinent to the concerns that have previously been discussed in this report. For example, studies of the reproductive outcomes of SAC crews might be informative. These crewmen, officers, and enlisted personnel have the same age distributions as the Atomic Veterans at the time of their exposure to ionizing radiation, a similar dose distribution, and access to uniform health care. Also, hospital records are available for most, if not all, births to the wives of these SAC crews. Several tens of thousands of individuals are involved. Presumably, much of the relevant information could be obtained by computer. Some of the potentially informative data are already available in a machine-retrievable format at the Armstrong Laboratory of Brooks Air Force Base in San Antonio, Texas. Moreover, a computer program, known as CARI, for estimating exposure to aircraft crews was developed at the request of the Federal Aviation Administration for commercial crews, but it should be applicable to Air Force crews as well. It requires information on flight pattern, duration, and altitude.

Such a study would not be without its limitations. These include the possible exposure to other potential mutagens such as chemicals, as well as heavy ion particles originating in space. Some of the information needed to estimate doses may be classified, but if the doses are computed by individuals in the Air Force with the appropriate security clearance, this might not be a major difficulty. It warrants noting, perhaps, that this same problem arose in the estimation of the doses received by the survivors of the atomic bombings of Hiroshima and Nagasaki and was overcome by using suitably cleared technical personnel.

Still other strategies, such as the applicability of recent developments in molecular and cellular biology, could be explored. These developments have given rise to a series of biological markers or biomarkers that can be used as estimators of exposure or dose, of biologic effects, and susceptibility. They include such chromosomal techniques as fluorescent in situ hybridization (FISH), and biochemical measures of damage to the genetic material (DNA) at specific loci, for example, the X-linked *hprt* locus or the autosomal glycophorin A locus. At present, these techniques have limited applicability as measures of exposure or dose to doses of less than 0.10 Gy (10 rad), and thus would seem to be of marginal use in the case of the Atomic Veterans, but this may change through the use of combined biomarker assays. The limitations of biomarkers as estimators of dose at doses of less than 0.10 Gy (10 rad) reflect to a substantial degree the relatively large, normally occurring interindividual response to a given dose. However, even now these markers could be useful as measures of effect or susceptibility. The strengths and current limitations of these methods have been addressed in recent workshops (NRC, 1995). It must be noted, however, that these techniques are often time-consuming, expensive, and difficult to implement with large numbers of individuals and cannot be routinely applied to all people. For example, the glycophorin A assay can be used only on individuals who are of the MN blood type, and these individuals constitute only half of most populations.

10

Conclusions

The committee explored in detail the feasibility of an epidemiologic study examining the association between adverse reproductive outcomes and paternal exposure to ionizing radiation. Such a study would be of great interest not only to the 210,000 veterans exposed to atomic weapons radiation but also to many other individuals who have received low doses of radiation at their places of employment or elsewhere. The committee's assessment is that it will be extremely difficult, if not impossible, to find and contact a sufficiently high and representative percentage of veterans' families, to establish a good measure of dose for each veteran, to identify and accurately document reproductive problems that occurred over a fifty-year interval, and to measure other factors that cause reproductive problems and therefore might confound any observed relationship between radiation exposure and reproductive problems. These difficulties become even more acute with regard to the grandchildren of these veterans. The cohort of Atomic Veterans does not provide a practical opportunity for a scientifically adequate and epidemiologically valid test of the hypothesis that paternal exposure to ionizing radiation has increased the frequency of adverse reproductive outcomes among their children and grandchildren. The committee recognizes the real concerns of the Atomic Veterans as expressed by their representatives, but it must conclude that epidemiologic studies cannot adequately address these concerns.

Glossary

Absorbed dose. When ionizing radiation passes through matter, some of its energy is imparted to the matter. The amount absorbed per unit mass of irradiated material is called the absorbed dose, and it is measured in gray or rad.

Aneuploid. A chromosome number that deviates from the normal for a species. It may be more or less than the diploid chromosome number but it is not an exact multiple of the basic haploid chromosome number found in germ cells.

Autosome. Chromosomes other than the sex (X or Y) chromosomes. Humans carry 22 pairs of autosomes.

Azospermia. Absence of sperm in a semen sample.

Background radiation. Radiation that is a natural part of a person's environment.

BEIR. Biological Effects of Ionizing Radiation. A series of reports by a committee of the National Academy of Sciences.

Beta particle. An electron with a positive or negative charge.

Carcinogenesis. The process of induction of cancer in a cell.

Case–control study. An epidemiologic investigation in which study subjects are selected on the basis of having a disease or condition and are compared in terms of a particular exposure to individuals who are selected because they do not have the condition. These studies are efficient for the study of rare diseases with long latency periods, but they are particularly susceptible to selection bias.

Chromosomes. Structural elements of various sizes in the cell nucleus composed of deoxyribonucleic acid (DNA) and proteins, they carry the genes that convey the genetic information of an organism. Chromosomes have a species-specific morphology and number.

Clusters (of adverse reproductive outcomes). Aggregation of events in space and time.

Cohort study. An epidemiologic investigation or follow-up of a group of individuals who are known to have had an exposure and are followed to see if they develop a disease or condition. Usually cohort studies are expensive and time-consuming because they typically follow a large number of individuals for many years.

Confidence Interval (CI). An interval within which a value for a population can lie with calculable probability (the probability that the true answer lies within the range). For example, an epidemiologic data point of 4.3 is subject to a number of uncertainties based on sample size and other factors. The value should not be assumed to represent the "truth" but rather an approximation of it. If a 95% CI of (2.3 to 7.8) is given for the data presented, this means that with repeated   analysis of this data the result will fall within this range 95% of the time.

Confounder. A variable that is causally related to the disease under study and is associated with the exposure in the study population, but is not a consequence of that exposure.

Congenital. Present at birth. Congenital does not imply either genetic or nongenetic causation.

Congenital malformation. Structural abnormality present at birth. This term is often used interchangeably with birth defect.

Contamination (radioactive). A radioactive substance in a material or place where it is undesirable.

Cosmic rays. Radiation of many sorts, but mostly atomic nuclei (protons) with very high energies originating outside the earth's atmosphere. Cosmic radiation is part of the natural background radiation. Some cosmic rays are more energetic than any synthetic forms of radiation.

Criticality accident. An unintentional event during which nuclear fuel (e.g., uranium-235) configures to initiate a chain reaction, releasing radiation and heat.

DDREF. A factor by which the effect caused by a specific dose of radiation changes at low dose rates compared with that at high dose rates.

Decay (radioactive). The spontaneous transformation of one nuclide (a nuclide is any atomic form of an element) into a different nuclide or into a different energy state of the same nuclide. The process results in a depletion, with time, of the radioactive atoms in a sample. Radioactive decay involves (1) the emission from the nucleus of alpha particles, beta particles (electrons), or gamma rays; (2) the nuclear capture or ejection of orbital electrons; or (3) fission. Also called radioactive disintegration. See Half-life.

Deformational. Subcategory of malformations in which the organ systems develop normally, but they incur a secondary deformity, for example, midline cleft palates.

Deletion. Loss of a portion of a chromosome as a result of chromosome break age.

Deoxyribonucleic acid (DNA). The long double-stranded molecule whose sequence, which consists of the four nucleotide bases (adenine, thymine, guanine, and cytosine), provides the genetic information of an organism.

DNA. See Deoxyribonucleic acid.

Dominant. A trait that is expressed in individuals who are heterozygous for a particular gene.

Dose (radiation). A term denoting the amount of energy absorbed. Absorbed dose is the energy imparted to matter by ionizing radiation for each unit of mass of irradiated material at the point of interest. It is usually expressed in rad (conventional units) or gray (international [SI] unit). Cumulative dose is the total dose resulting from repeated or continuous exposures to radiation.

Dose equivalent (H). A unit of biologically effective dose, defined by the ICRP in 1977 as the absorbed dose in rads multiplied by the quality factor (Q). For all X rays, gamma rays, beta particles, and positrons likely to be used in nuclear medicine, the quality factor is 1. The dose equivalent (H) is given by the equation $H = DQN$, where D is absorbed dose, Q is the quality factor, and N is the product of modifying factors (N is usually 1). See also Equivalent dose.

Dose rate (radiation). The radiation dose delivered for each unit of time and measured, for instance, in rad per hour. See Absorbed dose.

Doubling dose. The dose of radiation that, under a given set of conditions, will lead to an overall mutation frequency that is double the spontaneous frequency.

Down syndrome. A pattern of abnormalities related, in most cases, to the presence of an extra number 21 chromosome in all of the body's cells, giving them 47 chromosomes instead of the usual 46. People with Down syndrome have various degrees of mental retardation and often have congenital malformations of the heart.

DREF. see DDREF.

Effective dose. A quantity that takes into account the difference in sensitivity of various body tissues and the effectiveness of different forms of radiation. This quantity is used to obtain a uniform expression of risk for stochastic effects. It was been defined by the ICRP in 1990 as the sum of equivalent doses that have been multiplied by a tissue weighting factor (W_T). The unit of effective dose is the sievert or rem.

Embryo. An organism in the first stages of development. In humans, this is generally considered to be the period from the end of the second week through the eighth week of gestation.

Equivalent dose. A newer version of dose equivalent. It was defined by the ICRP in 1990 and uses radiation weighting factors (W_R) instead of the older quality factor (Q).

Euploid. A chromosome number that is the exact multiple of the haploid chromosome number.

Excess relative risk (ERR). Relative risk minus 1.

Exposure. A term relating, in this report, to the amount of ionizing radiation that is incident on living or inanimate material.

Exposure rate. Increment of exposure expressed for each unit of time.

F_1. In this report, first generation conceived after exposure of parent(s).

Fecundability. The ability to reproduce.

Fertility. The capacity to conceive or induce conception.

Fetus. Unborn offspring. In humans it is the period from 8 weeks after fertilization until birth.

Film badge. Photographic film shielded from light. It is worn by an individual to measure radiation exposure.

Gamete. A mature male or female germ cell (sperm or egg) containing a haploid number of chromosomes.

Gamma ray. Electromagnetic radiation emitted from the nucleus.

Gene. The basic unit of heredity. A finite segment of DNA that controls the production of a specific polypeptide.

Genetic code. The triplet sequence of nucleotide bases in the DNA chain, as reflected in messenger ribonucleic acid (mRNA), that determines the sequence of amino acids during protein synthesis.

Genetic effects of radiation. Radiation effects that can be transferred from parent to offspring. Any radiation-caused changes in the genetic material of germ cells. Compare to somatic effects of radiation.

Genetically significant dose (GSD). The GSD is that dose (of radiation) which, if received by every member of the population, would be expected to produce the same genetic injury to the population as that caused by the actual doses received by the individuals irradiated. The GSD is expressed in Sieverts (or rem).

Genotype. The genetic constitution of an individual that determines the physical and chemical characteristics of that individual.

Germ cell. Refers to male and female reproductive cells at their various levels of development (i.e., spermatogonia or oogonia, spermatocytes or oocytes and spermatozoon or ovum).

Gray (Gy). The international (SI) unit of radiation absorbed dose. One gray is equal to an energy deposition of 1 joule per kilogram (100 rad).

Half-life (radioactive). The time required for a radioactive substance to lose 50% of its activity by decay.

Heterozygous. Having dissimilar alleles at the same locus on homologous chromosomes.

High dose. More than 2 Gy (200 rad).

Homozygous. Having the same genes at a given locus on homologous chromosomes.

ICRP. International Commission on Radiological Protection.

Incidence. The number of people who have developed a disease in a given period of time divided by the total population at risk.

Induced burden. An increased number of genetic effects because of an exposure.

Intermediate dose. A dose of 0.2 to 2.0 Gy (20 to 200 rad).

Ionization. The process whereby a charged portion (usually an electron) of an atom or molecule is given enough kinetic energy to dissociate.

Ionizing radiation. Radiation that produces ion pairs along its path through a substance.

Irradiation. Exposure to radiation.

Isotopes. Nuclides with the same numbers of protons but different numbers of neutrons.

Klinefelter syndrome. A pattern of abnormalities related to the presence of an extra sex chromosome. Persons with Klinefelter syndrome are males, but they have 47 chromosomes, two X chromosomes and one Y chromosome, instead of the usual male chromosomal makeup of 46 chromosomes, with one X and one Y.

Latent period. Usually refers to the time elapsed between radiation exposure and the clinical appearance of an effect (such as the appearance of a cancer).

LET. See Linear energy transfer.

Linear energy transfer (LET). Amount of energy lost by ionizing radiation by way of interaction with matter for each unit of path length through the absorbing material.

Locus. The site occupied by a specific gene, or allele, on a particular chromosome.

Low birth weight. A weight of less than 2,500 grams at birth.

Low dose. A dose of less than 0.2 Gy (20 rad).

Low dose rate. Less than 0.1 mGy (100 mrad) per minute averaged over about 1 hour. It sometimes also refers to less than 10 mGy (1 rad) per year.

Micro- (μ). A prefix that divides a basic unit by 1 million.

Milli- (m). A prefix that divides a basic unit by 1,000.

Multifactorial. Refers to causation involving the interaction of a number of factors, often including several genes and nongenetic (environmental) factors.

Mutation. A hereditary change in genetic material. A mutation can be a change in a single gene (point mutation) or a change in the order or number genes.

Mutation rate. The rate at which mutations occur at a given locus, expressed as the number of mutations per gamete for each locus in a generation.

NCRP. National Council on Radiation Protection and Measurements.

Nuclear Regulatory Commission (NRC). A U.S. government agency regulating by-product material.

Nucleus (atom). The small, positively charged core of an atom. It is only about 1/10,000th of the diameter of the atom, but, it contains nearly all of the atom's mass. All nuclei, except the nucleus of ordinary hydrogen, contain both protons and neutrons, the nucleus of ordinary hydrogen, consists of a single proton.

Nucleus (cell). A mass of protoplasm within the cytoplasm of a cell. It is surrounded by a membrane and contains substances that direct the cell's me tabolism, growth, and reproduction.

Nuclides. A general term applicable to all atomic forms of an element. The term is often used incorrectly as a synonym for isotope, which properly has a more limited definition. Whereas isotopes are the various forms of a single element (hence, are a family of nuclides) and all have the same atomic number and number of protons, nuclides comprise all of the isotopic forms of all of the elements. Nuclides are distinguished by their atomic number, atomic mass, and energy state.

Odds ratio (OR). Used as an estimation of relative risk. It is primarily used for case–control studies and is calculated from the odds of exposure among the cases to that among controls.

Oligospermic. A reduced sperm count; it has various in definitions but often refers to a count of less than 20 million sperm per milliliter of semen.

Preeclampsia. Hypertension induced by pregnancy after 20 weeks of gestation that is accompanied by proteinuria, edema or both.

Preterm birth. A birth that occurs at a gestational age of less than 37 completed weeks (<259 days).

Prevalence. The proportion of individuals in a population who have a disease at a specific time, for example, the number of people in the United States who have lung cancer. It is not the number of new cases in that year. Prevalence is the incidence multiplied by the average duration of disease.

Quality factor. (QF). Dependent factor by which absorbed doses are to be multiplied to account for the various degrees of effectiveness of different radiations. QF for 250-kVp X rays is equal to 1. See also Weighting factor (radiation).

Rad. Radiation absorbed dose. A unit of absorbed dose of ionizing radiation. One rad is equal to 100 ergs/g. See Gray.

Radiation. Energy propagated through space or matter as waves (gamma rays, ultraviolet light) or as particles (alpha or beta particles). External radiation is from a source outside the body, whereas internal radiation is from a source inside the body (such as radionuclides deposited in tissues).

Radiation therapy. Treatment of disease with any type of radiation. Often called radiotherapy.

Radioactivity. The property of some nuclides of spontaneously emitting radiation.

Radionuclide. Unstable nucleus that transmutes by way of nuclear decay.

Radiosensitivity. A relative susceptibility of cells, tissues, organs, or organisms to the harmful action of radiation.

Recessive. Refers to a gene that produces its effect (is expressed) only when it is present in the homozygous or hemizygous state.

Relative risk (RR). The ratio of the disease incidence in the exposed population divided by the incidence in the nonexposed population. If there is no differ ence as a result of exposure, the RR is 1.0. A relative risk of 1.1 indicates a 10% increase in the number of cases than would be expected.

Rem. See Roentgen equivalent man.

Risk, absolute. In this report, the excess risk attributed to irradiation and usually expressed as the numeric difference between irradiated and nonirradiated populations (e.g., one excess case of cancer/1 million people irradiated an nually for each rad). Absolute risk may be given on an annual or lifetime (70-year) basis.

Roentgen (R). Quantity of X- or gamma-ray radiation per cubic centimeter of air that produces one electrostatic unit of charge.

Roentgen equivalent man (rem). The unit of the biologically effective dose. The absorbed dose in rad multiplied by the quality factor of the type of radia tion. See Sievert.

Sex chromosomes. Chromosomes that determine the sex of the individual: in humans, the X chromosome in the female, the X and Y chromosomes in the male.

Sex-linked gene. A gene located on a sex chromosome. Often used to describe genes on the X chromosome, although X-linked is the more accurate term.

Sievert (Sv). The international (SI) unit of dose equivalent. The absorbed dose in gray multiplied by the quality factor or radiation weighting factor of the type of radiation. One sievert equals 100 rem.

Somatic cells. All cells in the body except gametes and their precursors.

Somatic effects of radiation. Effects of radiation that are limited to the exposed individual, as distinguished from genetic effects, which only affect subse- quent unexposed generations. Large radiation doses can cause somatic ef- fects that are fatal. Lower doses may make the individual noticeably ill, may produce temporary changes in blood cell levels detectable only in the laboratory, or may have no detectable effect.

Somatic mutation. A mutation occurring in a somatic cell (i.e., one that is not passed on to future generations).

Spontaneous abortion. Non-deliberate interruption of intra-uterine pregnancy before 28 of weeks gestation in which the embryo or fetus is dead when delivered.

Stillbirth. Fetal death at 28 weeks of gestation or later (from last menstrual period).

Stochastic effect. An effect whose probability of occurrence in an irradiated population or individual is a function of dose. Commonly regarded as having no threshold dose. An example is radiation carcinogenesis.

Teratogenic. Related to the induction of structural malformations in an embryo and fetus.

Teratogens. Agents that can cause structural malformations in the developing embryo or fetus.

Threshold dose. The minimum dose of radiation that will produce a detectable biologic effect.

Tissue weighting factor. See weighting factor.

Toxicant. The actions and effects of a toxicant are similar to those of a toxin, but the source is a synthetic or artificial substance rather than a natural one.

Toxin. A natural substance that can induce a poisonous effect.

Translocation. The transfer of genetic material from one chromosome to another, nonhomologous chromosome. An exchange of genetic material between two chromosomes, each of which retains a centromere, is referred to as a reciprocal translocation. When a small fragment, which is usually lost, is formed (centric fusion), this is referred to as a Robertsonian translocation.

Triplet. In molecular genetics, a unit of three successive bases in DNA or RNA, coding for a specific amino acid.

Trisomy. A state in which there are three members of a given chromosome instead of the normal pair.

Turner syndrome. A pattern of abnormalities related to the absence of a sex chromosome. People with Turner syndrome have 45 chromosomes rather than the usual number of 46 and are female. They have a single X chromosome, and the syndrome is often referred to as XO Turner syndrome.

UNSCEAR. United Nations Scientific Committee on the Effects of Atomic Radiation. A series of reports by a committee of the United Nations.

Weighting factor (W$_r$). A number of values that are used to adjust for the various sensitivities of tissues and effectiveness of radiations in order to express risk. The values listed here were derived by the International Commission on Radiological Protection in 1990 and refer to stochastic effects only.

Weighting factor (radiation)

photons	1.0
electrons and muons	1.0
neutrons	5–20 (the value depends upon the energy)
protons	5.0
alpha particles	20.0

Weighting factors (tissue)

bone	0.01
skin	0.01
bladder	0.05
breast	0.05
liver	0.05
esophagus	0.05
thyroid	0.05
remaining organs	0.05
bone marrow	0.12
colon	0.12
lung	0.12
stomach	0.12
gonads	0.20

X chromosomes. A sex chromosome found in duplicate in the normal female and singly in the normal male.

X-linked. Genes carried on the X chromosome.

Y chromosome. One of the sex chromosomes found in the normal male. The Y chromosome is essential for the development of male gonads.

References

ABCC (Atomic Bomb Casualty Commission): Report of the Committee for Scientific Review of ABCC, February 1975. Atomic Bomb Casualty Commission Technical Report 21-75. 1975.

Andrews, G. A., K. F. Hubner, S. A. Fry, et al. Report of 21 year medical follow-up of survivors of the Oak Ridge Y-12 accident. In The Medical Basis of Radiation Accidents Preparedness. New York: Elsevier-North Holland, 1980.

Arakawa, E. T. Technical Report of Residual Radiation in Hiroshima and Nagasaki. Atomic Bomb Casualty Commission Technical Report 2-62. Hiroshima, Japan. 1962.

Awa, A. A., T. Honda, and S. Neriishi. Cytogenetic studies of the offspring of atomic bomb survivors. In Cytogenetics: Basic and Applied Aspects. G. Obe and A. Basler (eds.). Berlin: Springer-Verlag, pp. 116–183, 1987.

Awa, A. A., T. Honda, and S. Neriishi. Cytogenetic study of the offspring of atomic bomb survivors, Hiroshima and Nagasaki. Radiation Effects Research Foundation Technical Report 21-88. Hiroshima, Japan. Radiation Effects Research Foundation, 1988.

Baird, P.A., T.W. Anderson, H.B. Newcombe, et al. Genetic Disorders in Children and Young Adults: A Population Study. American Journal of Human Genetics. 42: 677-693, 1988.

Berkowitz, G. S. Environmental and occupational hazards to pregnancy. In Rovinsky and Guttmacher's Medical, Surgical and Gynecological Complications of Pregnancy. S. H. Cherry, G.S. Berkowitz, and N. G. Kase (eds.). Baltimore: Williams & Wilkins, 1985.

Berkowitz, G. S., and M. Marcus. Occupational exposures and reproduction. In Current Obstetric Medicine, Vol. 2. St. Louis: Mosby-Year Book Inc., pp. 54, 1993.

Berkowitz, G. S., and E. Papiernik. Epidemiology of preterm birth. Epidemiologic Reviews 15:414–443, 1993.

Biava, C., E. Smucklerm, and D. Whorton. The testicular morphology of individuals exposed to dibromochloropropane. Experimental and Molecular Pathology 29:448–458, 1978.

Bithell, J. F., and A. M. Stewart. Pre-natal irradiation and childhood malignancy: A review of British data from the Oxford survey. British Journal of Cancer. 31:271–287, 1975.

Bonde, J. P. Subfertility in relation to welding. Danish Medical Bulletin 37:105–108, 1990.

Bracken, M. B. The epidemiology of perinatal disorders. In Principles and Practice of Perinatal Medicine. J. B. Warshaw and J. C. Hobbins (eds.). Menlo Park, Calif.: Addison-Wesley, 1983.

Bracken, M. B. (ed.). Perinatal Epidemiology. New York: Oxford University Press, 1984.

Brown, N. A. Are offspring at risk from their father's exposure to toxins? Nature 316:110, 1985.

Caldwell, G. G. Twenty-two years of cancer cluster investigations at the Centers for Disease Control. American Journal of Epidemiology 132:S43–S47, 1990.

Caldwell, G. G., D. B. Kelley, and C. W. Heath. Leukemia among participants in military maneuvers at a nuclear bomb test. A preliminary report. Journal of the American Medical Association 244:1575–1578, 1980.

Caldwell, G. G., D. B. Kelley, and M. Zack. Mortality and cancer frequency among military nuclear test (SMOKY) participants, 1957 through 1979. Journal of the American Medical Association 250:620–624, 1983.

Carr, D. H., and M. Gedeon. Population cytogenetics of human abortuses. In Population Cytogenetics, Studies in Humans. E. B. Hook and I. H. Porter (eds.). New York: Academic Press, pp. 1–9, 1977.

Carter, C. O. Monogenetic disorders. Journal of Medical Genetics 14:316–320, 1977.

Centers for Disease Control. Guidelines for investigating clusters of health events. Morbidity and Mortality Weekly Report 39 (RR-11):1–23, 1990.

Childs, J. D. The effect of a change in mutation rate on the incidence of dominant and X-linked recessive disorders in man. Mutation Research 83:145–158, 1981.

Cook-Mozaffari, P., S. Darby, and R. Doll. Cancer near potential sites of nuclear installations. Lancet ii:1145–1147, 1989.

Crow, J. F., and C. Denniston. The mutation component of genetic damage. Science 212:888–893, 1981.

Czeizel, A., and K. Sankaranarayanan. The load of genetic and partially genetic disorders in man. I. Congenital anomalies: Estimates of detriment in terms of years of life lost and years of impaired life. Mutation Research 128:73–103, 1984.

Deeg, H., R. Storb, and E. Thomas. Bone marrow transplantation: A review of delayed complications. British Journal of Haematology 57:185–208, 1984.

Denniston, C. Low level radiation and risk estimation in man. Annals of Reviews in Genetics 16:329–355, 1982.

DHEW (Department of Health, Education and Welfare). The Collaborative Perinatal Study of the National Institute of Neurological Diseases and Stroke: The Women and Their Pregnancies. DHEW Publication No. (NIH) 73-379, Washington, D.C.: U.S. Government Printing Office, 1973.

DHHS (Department of Health and Human Services): Monthly Vital Statistics Report. Vol. 43, No. 6, Suppl. E. Washington, D.C.: U.S. Government Printing Office, 1995.

DNA (Defense Nuclear Agency). Dose binning report (unpublished data), March 8, 1995a.

DNA (Defense Nuclear Agency). Dose binning report (unpublished data), March 17, 1995b.

Doucette, J.T., and M. B. Bracken. Possible role of asthma in the risk of preterm labor and delivery. Epidemiology. 4 (2):143–150, 1993.

Ehling, U.H. Genetic Risk Assessment. Annals of Reviews in Genetics. 25:255–280, 1991.

Gardner, M. J. Paternal occupations of children with leukemia (letter). British Medical Journal. 305:715, 1992.

Gardner, M. J., M. P. Snee, and A. J. Hall. Results of case-control study of leukaemia and lymphoma among young people near Sellafield nuclear plant in West Cumbria. British Medical Journal 300:423–429, 1990a.

Gardner, M. J., A. J. Hall, and M. P. Snee. Methods and basic data of case-control study of leukaemia and lymphoma among young people near Sellafield nuclear plant in West Cumbria. British Medical Journal 300:429–434, 1990b.

Geneseca, A., M. R. Cabalin, and R. Miro. Human sperm chromosomes: Long-term effect of cancer treatment. Cancer Genetics and Cytogenetics 46:251–260, 1990.

Gesell, A. L., and C. S. Amatruda. Developmental Diagnosis: The Evaluation and Management of Normal and Abnormal Neuropsychologic Development in Infancy and Early Childhood. 3rd ed. H. Knobloch and B. Pasamanick (eds.). New York: Harper and Row, 1975.

Glasstone, S. (ed.). The Effects of Nuclear Weapons. U.S. Department of Defense. Washington, D.C.: U.S. Atomic Energy Commission, 1962.

Goldstein, L., and D. P. Murphy. Etiology of the ill-health in children born after maternal pelvic irradiation. 2. Defective children born after post-conception pelvic irradiation. American Journal of Roentgenology 22:322-331, 1929.

Grossman, H. J. (ed.). Manual on Terminology and Classification in Mental Retardation. Washington, D.C.: American Association on Mental Deficiency, 1977.

Hartikainen-Sorri, A. L., and M. Sorri. Occupational and socio-medical factors in preterm birth. Obstetrics and Gynecology 74:13–16, 1989.

Hatch, M. and M. Marcus. Occupational exposures and reproduction. In Reproductive and Perinatal Epidemiology. M. Kiely (ed.) Boca Raton, Fl.,: CRC Press, pp. 131–142, 1991.

Hawkins, M. M. Is there evidence of a therapy-related increase in germ cell mutation among childhood cancer survivors? Journal of the National Cancer Institute 83:1642–1650, 1991.

Hill, A. B. The Environment and Disease: Association or Causation? Proceedings of the Royal Society of Medicine 58:295–300, 1965.

Hill, C. And A. Laplanche, A.: Overall Mortality and Cancer Mortality around French Nuclear Sites. Nature 347:755-757, 1990.

Hook, E. B. Human chromosome abnormalities. In Perinatal Epidemiology. M. B. Bracken (ed.). New York: Oxford University Press, 1984.

Hook, E. B., and J. L. Hamerton. The frequency of chromosomal abnormalities detected in consecutive newborn studies—differences between studies—results by sex and by severity of phenotypic involvement. Journal of Medical Genetics 14:63–79, 1977.

IAEA (International Atomic Energy Agency). Chernobyl Technical Report. Vienna, Austria: International Atomic Energy Agency, 1991.

ICRP (International Commission on Radiological Protection). 1990 Recommendations of the International Commission on Radiological Protection, ICRP Publication 60, Annals of the ICRP 21. New York: Pergamon Press, 1991.

IOM (Institute of Medicine). A review of the dosimetry data available in the Nuclear Test Personnel Review (NTPR) program. An interim letter report of the Committee to Study the Mortality of Military Personnel Present at Atmospheric Tests of Nuclear Weapons to the Defense Nuclear Agency. May 15, 1995.

Jablon, S., Z. Hrubec, and J. D. Boice. Cancer in populations living near nuclear facilities. A survey of mortality nationwide and incidence in two states. Journal of the American Medical Association 265:1403–1408, 1991.

Jacobs, P. A. The load due to chromosome abnormalities in man. In The Role of Natural Selection in Human Evolution. F. M. Salzano. (ed.) Amsterdam: North Holland, pp. 337–352, 1975.

Kato, H., W. J. Schull, and J. V. Neel. A cohort-type study of survival in the children of parents exposed to atomic bombings. American Journal of Human Genetics 16:214–230, 1966.

Kelsey, J. L., W. D. Thompson, and A. S. Evans. Methods in Observational Epidemiology. New York: Oxford University Press, 1986.

Knudson, A. G. Rare cancers: Clues to genetic mechanisms. Princess Takamatsu Symposia 18:221–231, 1988.

Levine, A. J. The genetic origins of neoplasia. Journal of the American Medical Association 273:592, 1995.

Li, F. P., J. F. Fraumeni, Jr., J. J. Mulvihill, et al. A cancer family syndrome in twenty-four kindreds. Cancer Research 48:5358–5362, 1988.

Lie, R. T., I. Huechand, and L. M. Irgens. Maximum likelihood estimation of the proportion of congenital malformations using double registration systems. Biometrics 50:433–444, 1994.

Little, M. P. A comparison between the risks of childhood leukaemia from parental exposure to radiation in the Sellafield work force and those displayed among the Japanese bomb survivors. Journal of Radiological Protection 10:185–198, 1990.

Little, M. P. A comparison of the apparent risks of childhood leukaemia from parental exposure to radiation in the 6 months prior to conception in the Sellafield workforce and the Japanese bomb survivors. Journal of Radiological Protection 11:77–90, 1991.

Little, M. P. The risks of leukaemia and non-cancer mortality in the offspring of the Japanese bomb survivors and a comparison of leukaemia risks with those in the offspring of the Sellafield workforce. Journal of Radiological Protection 12:203–218, 1992.

Lundsberg, L.S., M. B. Bracken, and K. Belanger. Occupationally related magnetic field exposure and male subfertility. Fertility and Sterility 63:384–391, 1995.

Lushbaugh, C. C. and G. W. Casarett. The effects of gonadal irradiation in clinical radiation therapy. Cancer 37:1111–1120, 1976.

Marcoux, S., J. Brisson, and J. Fabia. The effect of cigarette smoking on the risk of preeclampsia and gestational hypertension. American Journal of Epidemiology 130:950–957, 1989.

Martin, R. H., K. Hildebrand, and J. Yamamoto. An increased frequency of human sperm chromosomal abnormalities after radiotherapy. Mutation Research 174:219–225, 1986.

McClure, R. D. Topics in primary care medicine. Western Journal of Medicine 144:365–368, 1986.

McLaughlin, J. R., E. A. Clarke, and E. D. Nishri. Childhood leukemia in the vicinity of Canadian nuclear facilities. Cancer Causes Control 4:51–58, 1993.

Mellin, G. W. The frequency of birth defects. In Birth Defects. M. Fishbein (ed.). Philadelphia: J. B. Lippincott Co., pp. 1–17, 1963.

Mettler, F., and A. Upton. Medical Effects of Ionizing Radiation, Second Edition. Philadelphia: W.B. Saunders Co., pp. 90, 1995.

Michaelis, J., B. Kellerand, and G. Haaf. Incidence of childhood malignancies in the vicinity of (West) German nuclear power plants. Cancer Causes Control 3:255–263, 1992.

Miller, R. W. Delayed effects occurring within the first decade after exposure of young individuals to the Hiroshima atomic bomb. Pediatrics 18:1–18, 1956.

Miller, R. W. The discovery of human teratogens, carcinogens and mutagens: Lessons for the future. In Chemical Mutagens: Principles and Methods for Their Detection. A. Hollaender and F. J. de Serres (eds.). New York: Plenum Publishing Corp., pp. 101–126, 1978.

Murphy, D. P. Ovarian irradiation—its effect on the health of subsequent children: Review of the literature, experimental and clinical, with a report of 320 human pregnancies. Surgery, Gynecology and Obstetrics 47:201–215, 1928.

National Center for Health Statistics: Vital statistics of the United States, 1983 Life Tables. Volume II, Section 6. Hyattsville, MD.: U. S. Department of Health and Human Services, DHHS Publication No. (PHS) 86-1104, 1986.

NCRP (National Council on Radiation Protection and Measurements). Ionizing Radiation Exposure of the Population of the United States. NCRP Report No. 93, Bethesda, Md.: National Council on Radiation Protection and Measurement, 1987.

Neel, J. V., and S. E. Lewis. The comparative radiation genetics of humans and mice. Annual Review of Genetics 24:328–362, 1990.

Neel, J. V., and W. J. Schull. Studies on the potential effects of the atomic bombs. Acta Genetica 6:183–196, 1956a.

Neel, J. V., and W. J. Schull. The Effect of Exposure to the Atomic Bombs on Pregnancy Termination in Hiroshima and Nagasaki. Washington, D.C.: National Academy of Sciences–National Research Council, Publication No. 461, 1956b.

Neel, J. V., and W. J. Schull. (eds.). The Children of Atomic Bomb Survivors: A Genetic Study. Washington, D.C.: National Academy Press, 1991.

Neel, J. V., N. E. Morton, and W. J. Schull. The effect of exposure of parents to the atomic bombs on the first generation offspring in Hiroshima and Nagasaki (preliminary report). Japanese Journal of Genetics 28:211–218, 1953.

Neel, J. V., H. Kato and W. J. Schull. Mortality in the children of atomic bomb survivors and controls. Genetics 76:311–326, 1974.

Neel, J. V., W. J. Schull, and A. A. Awa. The children of parents exposed to atomic bombs: Estimates of the genetic doubling dose of radiation for humans. American Journal of Human Genetics 46:1053–1072, 1990.

Neutra, R. R., S. Swan, and D. Freedman. Clusters galore (abstract). Archives of Environmental Health 45:314, 1990.

NRC (National Research Council). Genetic effects of the atomic bomb in Hiroshima and Nagasaki. Genetic conference, Committee on Atomic Casualties. Science 106:331–333, 1947.

NRC (National Research Council). Review of the methods used to assign radiation doses to service personnel at nuclear weapon tests. Washington, D.C.: National Academy Press, 1985.

NRC (National Research Council). Committee on the Biological Effects of Ionizing Radiation. Health Effects of Exposure to Low Levels of Ionizing Radiation (BEIR V). Washington, D.C.: National Academy Press, 1990.

NRC (National Research Council). Committee on an Assessment of CDC Radiation Studies. Washington, D.C.: National Academy Press, 1995.

Nuclear Regulatory Commission. Health Effects Model for Nuclear Power Plant Accident Consequence Analysis. J. S. Evans, D. W. Moeller, and D. W. Cooper (eds.). NUREG/CR-4214. Washington, DC.: U.S. Nuclear Regulatory Commission, 1985.

Oftedal, P., and A. G. Searle. An overall genetic risk assessment for radiological protection purposes. Journal of Medical Genetics 17:15–20, 1980.

O'Leary, L. M., A. M. Hicks, and J. M. Peters. Paternal occupational exposures and risk of childhood cancer: A review. American Journal of Industrial Medicine 20:17–35, 1991.

Olshan, A. F., and E. M. Faustman. Male-mediated developmental toxicity. Annual Review of Public Health 14:159–181, 1993.

Olsnan, A. F., and P. Schnitzer. Paternal occupation and birth defects. In Male-Mediated Developmental Toxicity. A. F. Olshan and D. R. Mattison (eds.). New York: Plenum, 1994.

OPRR (Office for Protection from Research Risks, National Institutes of Health). Protecting Human Research Subjects: Institutional Review Board Guidebook. Washington, D.C.: U.S. Government Printing Office, pp. 4–10 to 4–14, 1993.

Ortin, T., C. Shostak, and S. Donaldson. Gonadal status and reproductive function following treatment for Hodgkin's disease in childhood. International Journal of Radiation Oncology, Biology Physics. 19(4):873–880, 1990.

OTA (Office of Technology Assessment). Infertility: Medical and Social Choices. OTA-BA-358. Washington, D.C.: U. S. Government Printing Office, p. 4, 1988.

Paterson, M. P., N.T. Hansen, P. J. Smith, and J. J. Mulvihill. Radiogenic neoplasia cellular radiosensitivity and faulty DNA repair. In Radiation Carcinogenesis: Epidemiology and Biological Significance. J. D. Boice, Jr., and J. F. Fraumeni, Jr. (eds.). New York: Raven Press, pp. 319–336, 1984.

Placek, P. J. Maternal and infant health factors associated with low infant birth weight: Findings from the 1972 National Natality Survey. In The Epidemiology of Prematurity. D. M. Reed and F. J. Stanley (eds.). Baltimore and Munich: Urban and Schwarzenberg, pp. 197–212, 1977.

Plummer, G. Anomalies occurring in children exposed in utero to the atomic bomb in Hiroshima. Pediatrics 10: 687–692, 1952.

Rawlings, J. S., V. B. Rawlings, and J. A. Read. Prevalence of low birth weight and preterm delivery in relation to the interval between pregnancies among white and black women. New England Journal of Medicine 332:69–74, 1995.

Robaire, B., and B. F. Hales. Post-testicular mechanisms of male-mediated developmental toxicity. In Male-Mediated Developmental Toxicity. A. F. Olshan and D. R. Mattison (eds.). New York: Plenum, 1994.

Rothenberg, R. B., K. K. Steinberg and S. B. Thacker. The public health importance of clusters: A note from the Centers for Disease Control. American Journal of Epidemiology 132:S3-5, 1990.

Russell, W. L. X-ray-induced mutations in mice. Cold Spring Harbor Symposia on Quantitative Biology 16:327–336, 1951.

Sanders, J., S. Pritchard, P. Mahoney, et al. Growth and development following marrow transplantation for leukemia. Blood 68:1128–1135, 1986.

Savitz, D. A. Childhood cancer. Occupational Medicine State of the Art Review 1:415–429, 1986.

Savitz, D. A., and J. Chen. Parental occupation and childhood cancer: A review of epidemiologic studies. Environmental Health Perspectives 88:325–337, 1990.

Savitz, D. A., N. Sonnenfeld, and A. F. Olshan. Review of epidemiologic studies of paternal exposure and spontaneous abortion. American Journal of Industrial Medicine 25:361–383, 1994.

Schull, W. J., and J. V. Neel. Radiation and the sex ratio in man. Science 128:343–348, 1958.

Schull, W. J., J. V. Neel, and A. Hashizume. Some further observations on the sex ratio among the offspring of survivors of the atomic bombings of Hiroshima and Nagasaki. American Journal of Human Genetics 18:328–336, 1966.

Schull, W., J. M. Otake, and J. V. Neel. Genetic effects of the atomic bombs: A reappraisal. Science 213:1220–1227, 1981a.

Schull, W. J., M. Otake, and J. V. Neel. Hiroshima and Nagasaki: A reassessment of the mutagenic effect of exposure to ionizing radiation. In Population and Biological Aspects of Human Mutation. E. B. Hook, and I. H. Porter (eds.). New York: Academic Press, pp. 277–303, 1981b.

Seibel, N. L., J. Cossman, and I. T. Magrath. Lymphoproliferative disorders. In Principles and Practice of Pediatric Oncology. 2nd ed. P. A. Pizzo, and D. G. Poplack (eds.). Philadelphia: J. B. Lippincott, Co. pp. 595–616, 1993.

Sever, L. E. The state of the art and current issues regarding reproductive outcomes potentially associated with environmental exposures: Reduced fertility, reproductive wastage, congenital malformations and birth weight. U.S. Environmental Protection Agency Workshop on Reproductive and Developmental Epidemiology: Issues and Recommendations. Report 600/8-89/103, Washington, D.C.: U.S. Environmental Protection Agency, 1989.

Shimizu, Y., H. Kato, and W. J. Schull. Life Span Study Report 11. 3. Non-cancer mortality, 1950–85, based on the revised doses (DS86). Radiation Research 130:249–266, 1992.

Shu, X.O., G. H. Reaman, and B. Lampkin. Association of paternal diagnostic X-ray exposure with risk of infant leukemia. Cancer Epidemiology, Biomarkers and Prevention 3:645–653, 1994.

Statistics and Epidemiology Research Corporation. EGRET SIZ: Sample Size and Power for Nonlinear Regression Models, Version 1. Seattle, Wash.: Statistics and Epidemiology Research Corporation, 1993.

Steeno, O. P. and A. Pangkahila. Occupational influences on male fertility and sexuality. Andrologia 16:93–101, 1984.

Stevenson, A. C. The load of hereditary defects in human populations. Radiation Research Supplement 1:306–325, 1959.

Swerdloff, R. S. Infertility in the male. Annals of Internal Medicine 103(6 pt 1): 906–919, 1985.

Teitelman, A. M., L. S. Welch, K. G. Hellenbrand, and M. B. Bracken. Effect of maternal activity on preterm birth and low birth weight. American Journal of Epidemiology 131(1): 104–113, 1990.

Tomatis, L. Transgenerational carcinogenesis: A review of the experimental and epidemiological evidence. Japanese Journal of Cancer Research 85:443-454, 1994.